Small Engine and Equipment Maintenance

Featuring Briggs & Stratton Powered Products

BRIGGS & STRATTON

atp AMERICAN TECHNICAL PUBLISHERS
ORLAND PARK, ILLINOIS 60467-5756

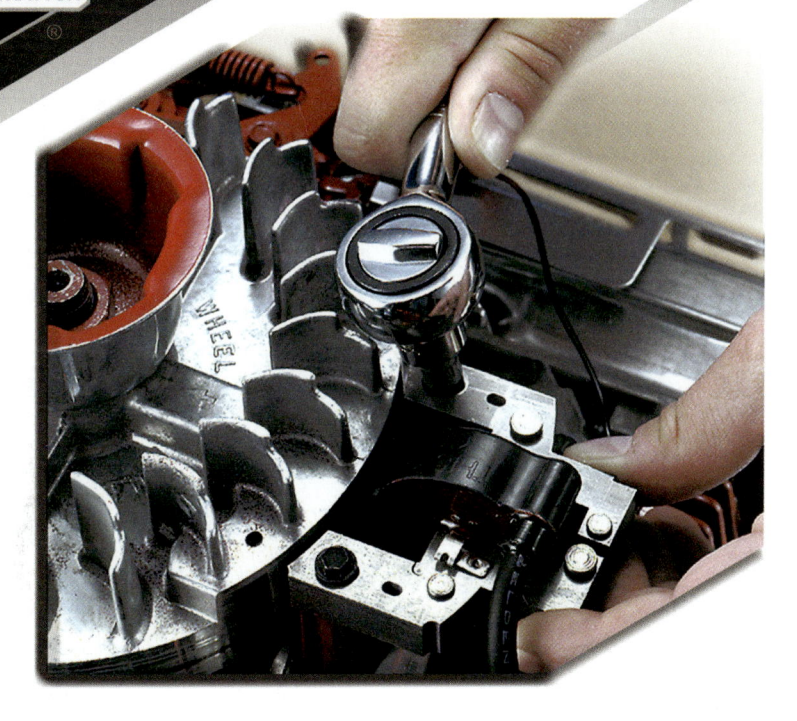

Donald R. Koloski

R. Bruce Radcliff

Small Engine and Equipment Maintenance contains procedures commonly practiced in industry and the trade. Specific procedures vary with each task and must be performed by a qualified person. For maximum safety, always refer to specific manufacturer recommendations, insurance regulations, specific job site and plant procedures, applicable federal, state, and local regulations, and any authority having jurisdiction. The material contained is intended to be an educational resource for the user. American Technical Publishers, Inc. assumes no responsibility or liability in connection with this material or its use by any individual or organization.

Acknowledgments

The author and publisher are grateful to the following organization for providing information and technical assistance.

Fluke Corporation
The Stanley Works
L.S. Starrett Company
Wacker Neuson Corporation
Bacharach, Inc.

American Technical Publishers, Inc., Editorial Staff

Editor in Chief:
    Jonathan F. Gosse
Vice President—Production:
    Peter A. Zurlis
Art Manager:
    James M. Clarke
Multimedia Manager:
    Carl R. Hansen
Technical Editor:
    James T. Gresens
Copy Editor:
    Talia J. Lambarki
Cover Design:
    Mark S. Maxwell

Illustration/Layout:
    Mark S. Maxwell
    Nick W. Basham
    James M. Clarke
    Melanie G. Doornbos
    Thomas E. Zabinski
    Samuel T. Tucker
CD-ROM Development:
    Gretje Dahl
    Dan Kundrat
    Nicole S. Polak
    Bonnie M. Rajchel

Microsoft, Windows, Windows Vista, PowerPoint, and Internet Explorer are either registered trademarks or trademarks of Microsoft Corporation in the United States and/or other countries. Adobe, Acrobat, and Reader are registered trademarks of Adobe Systems Incorporated in the United States and/ or other countries. Intel is a registered trademark of Intel Corporation in the United States and/or other countries. Firefox is a registered trademark of the Mozilla Foundation. Quantum, Ninja, and Intek are either registered trademarks or trademarks of Briggs & Stratton Corp. Cobalite is a registered trademark of Umicore. National Electrical Code and NEC are registered trademarks of the National Fire Protection Association, Inc. Torx is a registered trademark of Camcar Corp., division of Textron Ind. UL is a registered trademark of Underwriters Laboratories. Vise-Grip is a registered trademark of Petersen Manufacturing Company, Inc. Weed Eater is a registered trademark of Husqvarna Consumer Outdoor Products (HCOP). Quick Quiz and Quick Quizzes are either registered trademarks or trademarks of American Technical Publishers, Inc.

© 2012 by American Technical Publishers, Inc.
All rights reserved

1 2 3 4 5 6 7 8 9 – 12 – 9 8 7 6 5 4 3 2 1

Printed in the United States of America

ISBN 978-0-8269-0044-9

This book is printed on recycled paper.

Contents

1 TOOLS AND SAFETY — 1
Tools • Safety

2 HISTORY OF SMALL ENGINES AND RELATED APPLICATIONS — 13
History of Outdoor Power Equipment

3 OUTDOOR POWER EQUIPMENT APPLICATIONS — 23
Outdoor Power Equipment Applications

4 SMALL ENGINE FUNDAMENTALS — 37
Four-Stroke Cycle Theory • Small Engine Systems

5 OUTDOOR POWER EQUIPMENT DEVICES — 63
Mechanical Drive Systems • Mechanical Switches • Implements and Attachments • OPE Maintenance • Storage

6 BASIC SMALL ENGINE MAINTENANCE AND REPAIR PROJECTS — 81
Changing Engine Oil • Removing Debris from Engines • Servicing Ignition Systems • Servicing Air Cleaners • Inspecting and Replacing Mufflers • Replacing Mower-Deck Drive Belts • Replacing Snow Thrower Shear Pins • Replacing Snow Thrower Skid Shoes • Engine Troubleshooting and General Maintenance • Preparing Equipment for Long-Term Shortage

7 INTERMEDIATE SMALL ENGINE MAINTENANCE AND REPAIR PROJECTS — 105
Overhauling Carburetors • Replacing Flywheels and Flywheel Keys • Servicing Flywheel Brakes • Servicing Fuel Systems • Servicing Governor Systems • Servicing Rewind Starters •

8 ADVANCED SMALL ENGINE MAINTENANCE AND REPAIR PROJECTS — 139
Replacing Drive Discs • Removing Carbon Deposits • Servicing Electrical Systems • Replacing Ignition Systems • Servicing Valves

APPENDIX — 175

GLOSSARY — 195

INDEX — 199

Introduction

Small Engine and Equipment Maintenance is a comprehensive full-color book that provides information on the troubleshooting and repair of outdoor power equipment including lawn mowers, lawn and garden tractors, snow throwers, pressure washers, portable generators, tillers, and wood chippers. The most common residential and light commercial small engine and outdoor power equipment systems are covered, including the compression, fuel, governor, electrical, cooling, and lubrication systems.

Small Engine and Equipment Maintenance is designed for outdoor power equipment users and for students in power technology, automotive, and small engine training programs. Fundamental small engine operation and application principles are presented using concise text, detailed illustrations, and informative factoids. Information is supplemented with photographs from recognized outdoor power equipment manufacturers. Projects for basic, intermediate, and advanced maintenance and repair tasks are included. All key terms are italicized in the chapters and defined in the Glossary, and the Appendix contains many useful charts and tables.

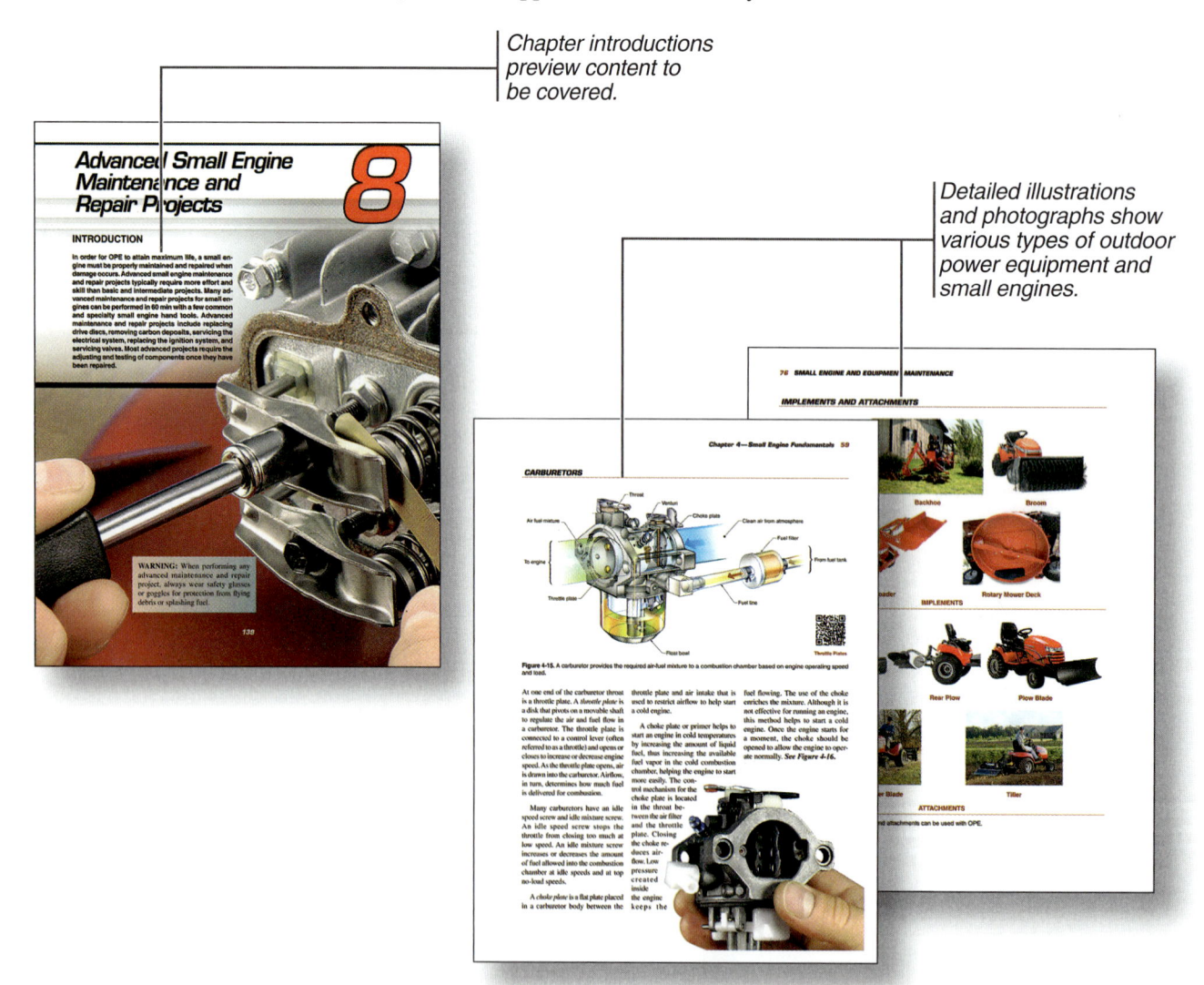

Chapter introductions preview content to be covered.

Detailed illustrations and photographs show various types of outdoor power equipment and small engines.

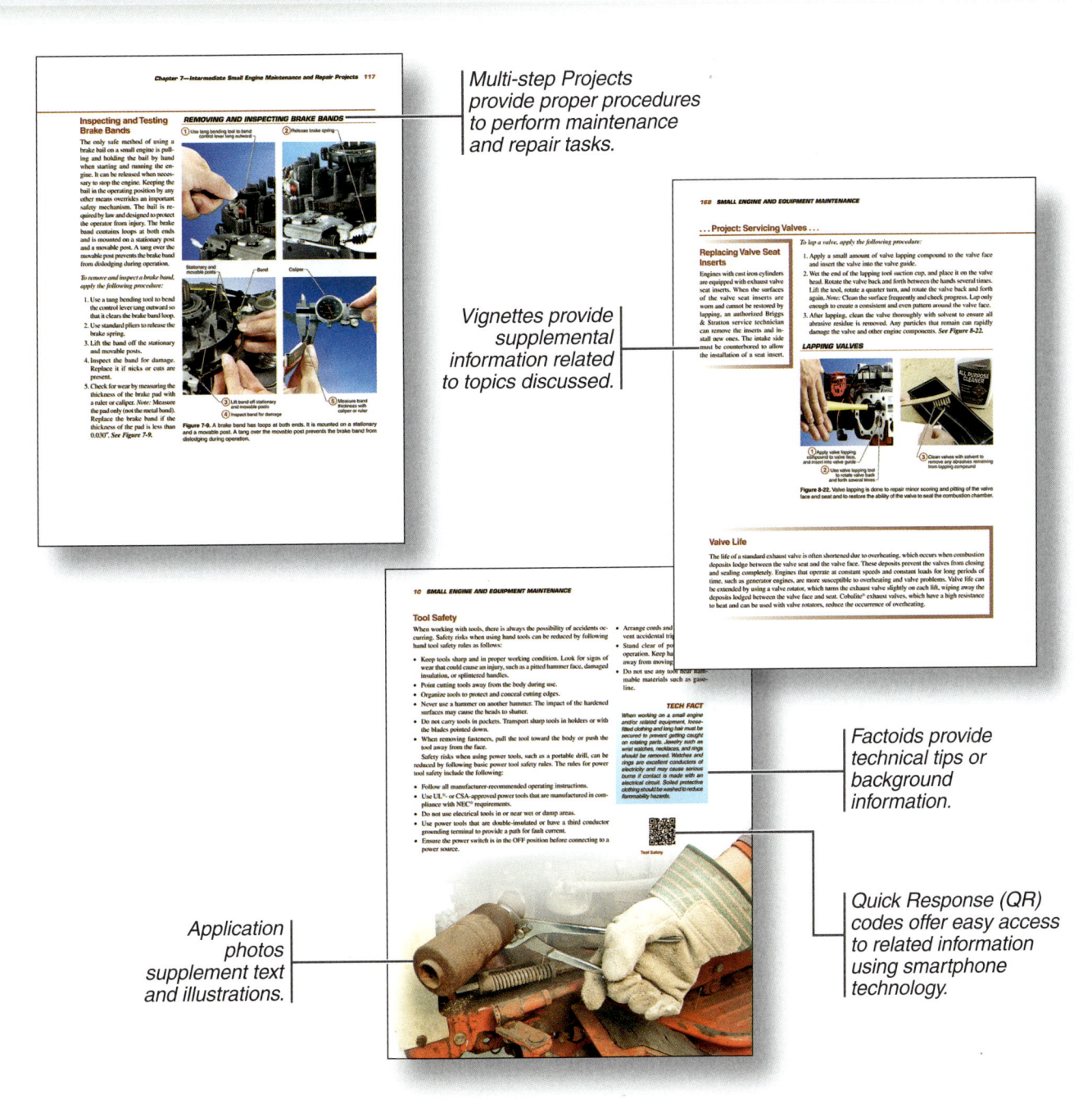

Multi-step Projects provide proper procedures to perform maintenance and repair tasks.

Vignettes provide supplemental information related to topics discussed.

Factoids provide technical tips or background information.

Quick Response (QR) codes offer easy access to related information using smartphone technology.

Application photos supplement text and illustrations.

To obtain information on related training products, visit the American Technical Publishers website at www.go2atp.com.

The Publisher

Tools and Safety

INTRODUCTION

Small engines are used to operate most outdoor power equipment, such as lawn mowers, snow blowers/throwers, chain saws, log splitters, rototillers, leaf blowers, weed trimmers, and pressure washers. The use of outdoor power equipment reduces user fatigue and allows easy performance of most tasks. However, small engine operation and the maintenance and repair of small engines involve the risk of injury. These risks may be reduced by following general safety precautions and performing tasks in a safe environment. Some tasks may require special tools and safety procedures.

TOOLS

The proper tools allow maintenance and repair tasks to be performed easily and provide the best results. Tools that should be available when performing work on small engines include general-purpose tools and the tools used specifically for small engines.

General-Purpose Tools

General-purpose tools are used for most small engine and outdoor power equipment maintenance and repair projects. Work cannot be performed without these tools. *See Figure 1-1.* General-purpose tools include the following hand and power tools:

- **adjustable pliers:** used to grip large fasteners and parts
- **adjustable wrenches:** used to loosen/tighten nuts and bolts in tight locations
- **basters:** used to remove oil or fuel from tanks that are not fitted with drain plugs
- **center punches:** used to mark center points for drilling holes
- **combination wrench sets:** used to loosen/tighten nuts and bolts with the convenience of an open end and box end on the same wrench
- **diagonal wire cutters:** used to cut wire and other materials of small diameter
- **flat files:** used to form or smooth rough metal edges
- **flathead screwdrivers:** used to drive slotted head fasteners
- **hacksaws:** used to cut metal; adjustable steel frames hold blades of various lengths and types; blades are inserted with the teeth pointed away from the handle

2 SMALL ENGINE AND EQUIPMENT MAINTENANCE

- **hex wrench sets (Allen wrenches):** used to loosen/tighten hex-head screws
- **locking pliers (Vise-Grip® pliers):** used to lock fasteners and/or hold parts
- **measuring rulers:** used to measure linear distance
- **needle-nose pliers:** used to grip small fasteners and/or parts in tight locations
- **parts cleaning brushes:** used to safely clean dirt and debris from small parts prior to installation in engines
- **Phillips-head screwdrivers:** used to drive Phillips-head fasteners

GENERAL-PURPOSE TOOLS

The Stanley Works
ADJUSTABLE PLIERS

The Stanley Works
ADJUSTABLE WRENCH

BASTER

The Stanley Works
CENTER PUNCHES

The Stanley Works
COMBINATION WRENCH SET

The Stanley Works
DIAGONAL WIRE CUTTER

The Stanley Works
FLAT FILE

Phillips-head
The Stanley Works
Flathead
SCREWDRIVERS

The Stanley Works
HACKSAW

The Stanley Works
HEX WRENCH SET

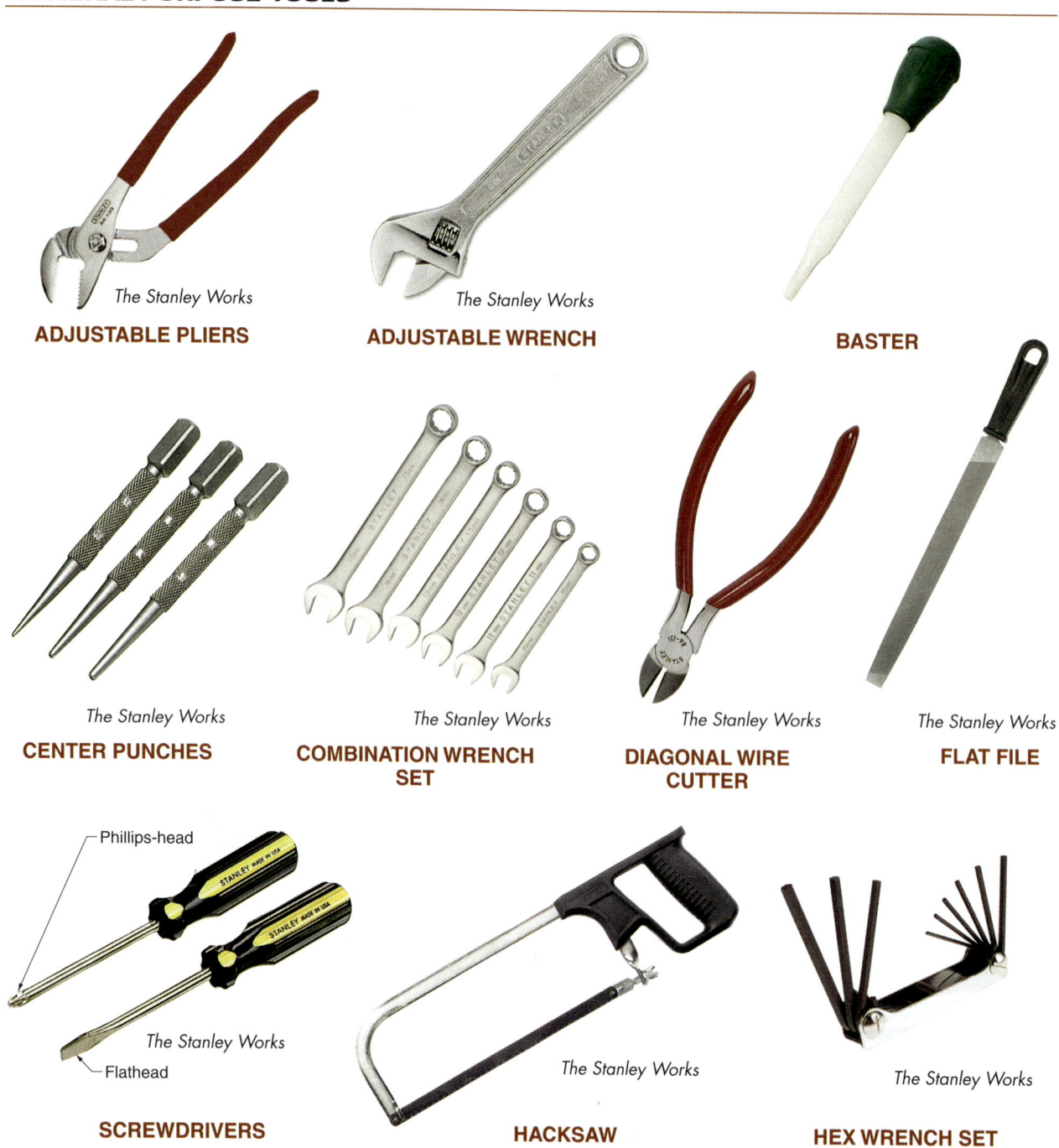

Figure 1-1. General-purpose tools are used for most maintenance and repair projects on small engines and outdoor power equipment.

Chapter 1—Tools and Safety 3

- **power drills:** used to drill holes in material or attach threaded fasteners
- **putty knives:** used to smooth materials (blade widths vary)
- **ratchet wrenches and socket sets:** used to loosen/tighten nuts and bolts in tight locations
- **shot-filled mallets:** used to hammer/drive material and parts without damaging surfaces; rubber mallets may also be used for this purpose
- **standard pliers (slip-joint pliers):** used to grip fasteners and parts
- **star-shaped (Torx®) driver sets:** used to drive Torx®-head fasteners

The Stanley Works
LOCKING PLIERS

L.S. Starrett Company
MEASURING RULERS

The Stanley Works
NEEDLE-NOSE PLIERS

PARTS CLEANING BRUSHES

The Stanley Works
STANDARD PLIERS

The Stanley Works
PUTTY KNIFE

The Stanley Works
SHOT-FILLED MALLET

The Stanley Works
RATCHET WRENCH AND SOCKET SET

STAR-SHAPED TORX® DRIVER SET

POWER DRILL (CORDLESS)

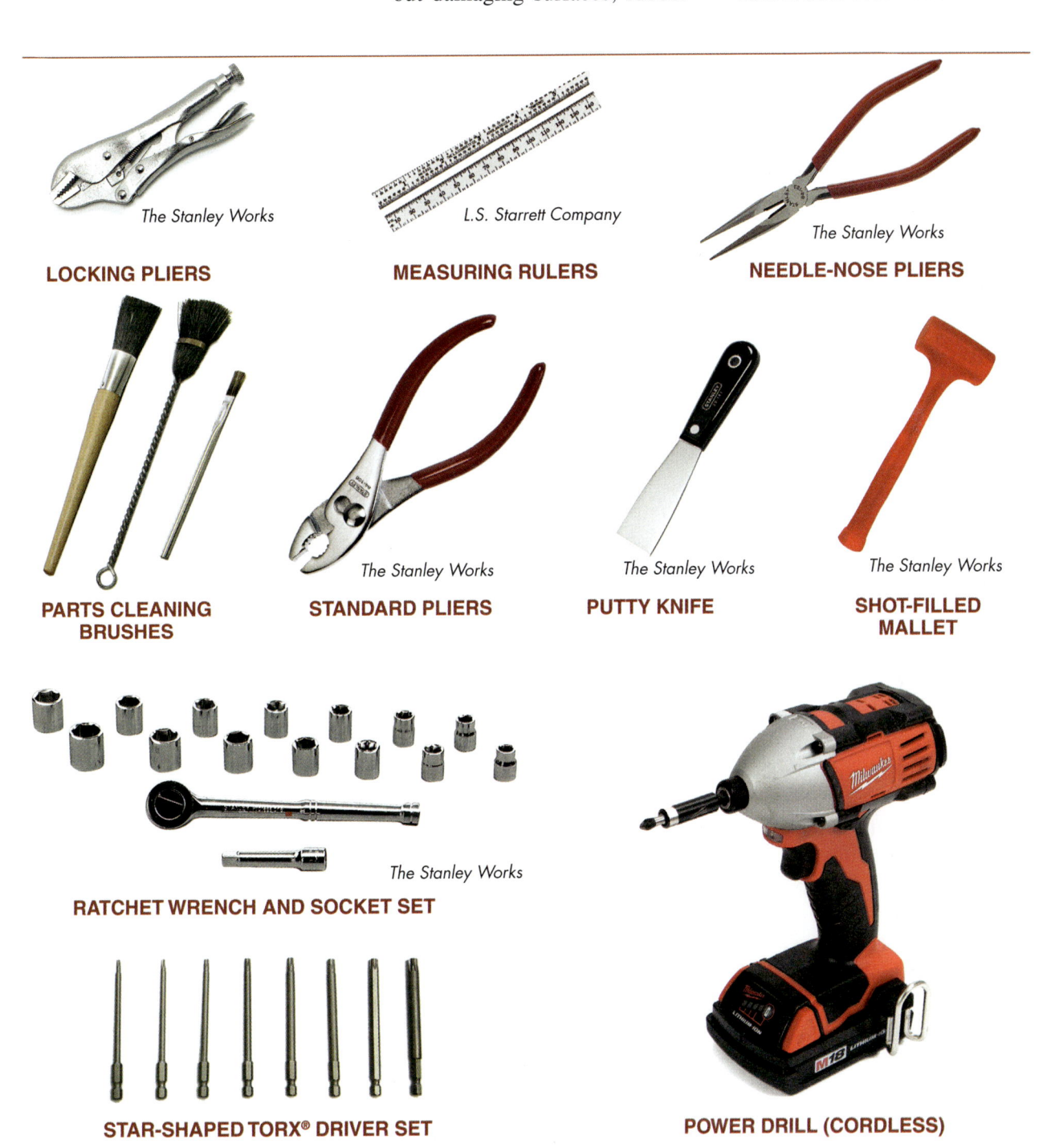

Figure 1-1. (Continued)

Small Engine Tools

Some tools are used specifically for performing maintenance and repair tasks on small engines and related equipment. These tools, while easy to use, are required for intermediate to advanced tasks. *See Figure 1-2.* Also, the correct tools should be used for each task. Damage to the engine can be caused and safety hazards created by using other tools as substitutes. Small engine specialty tools include the following:

- **multimeters:** used to take electrical measurements such as voltage, current, and resistance
- **feeler gauges:** used to measure the distance between tightly fitted parts
- **flywheel holders:** used to hold flywheels in place when removing or installing flywheel nuts or rewind starter clutches
- **flywheel pullers:** used to remove flywheels
- **fuel line clamp tools:** used to attach to engines
- **oil evacuator pumps:** used to pump oil or fuel from tanks that are not fitted with drain plugs
- **torque wrenches:** used to tighten nuts and bolts to a specific torque level
- **spark plug gauges:** used to measure the gap at the tip of spark plugs
- **spark testers:** used to test the condition of ignition systems by testing spark plug gap
- **starter clutch wrenches:** used to remove and install rewind starter clutches
- **tachometers:** used to measure the rotational speeds of objects; expressed as revolutions per minute (RPM); modern tachometers are attached to the housing of running machines rather than being in contact with rotating shafts
- **tang benders (governor tang benders):** used to adjust governor

SMALL ENGINE TOOLS

MULTIMETERS
Digital (DMM)
Analog
Fluke Corporation

FEELER GAUGES

FLYWHEEL HOLDER

FLYWHEEL PULLER

FUEL LINE CLAMP TOOL

OIL EVACUATOR PUMP

TORQUE WRENCHES
Beam type
Audible type

Figure 1-2. Some tools are used specifically for performing maintenance and repair tasks on small engines and related equipment and are required for intermediate to advanced level tasks.

Digital Multimeters

Chapter 1—Tools and Safety 5

tabs for top no-load speed without the use of pliers
- **valve spring compressors:** used to compress valve springs

High-quality tools should be used for small engine repair. Well-made tools perform better and last longer than those poorly made.

High-quality tools are identified by their sturdy construction and ability to provide a tight fit or grip on any type of workpiece.

. . . SMALL ENGINE TOOLS

SPARK PLUG GAUGE

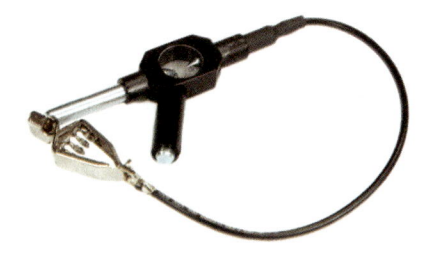

SPARK TESTER

STARTER CLUTCH WRENCH

TACHOMETER

TANG BENDER

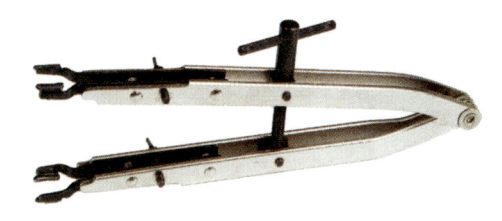

VALVE SPRING COMPRESSOR

Figure 1-2. (Continued)

6 SMALL ENGINE AND EQUIPMENT MAINTENANCE

SAFETY

Safety hazards must be avoided at all times such as exposure to exhaust fumes. For example, small engine exhaust fumes contain carbon monoxide. *Carbon monoxide (CO)* is an odorless, colorless, and poisonous gas produced by burning gasoline and other fuels. A carbon monoxide meter or gas meter can alert individuals in the area of the presence of CO indoors before it reaches lethal levels. *See Figure 1-3.* Carbon monoxide accumulation can be avoided by working outdoors or in well-ventilated areas.

Since small engines burn fuel and induce electricity, tasks involving small engines require special safety considerations. The safety considerations for small engines apply to operating conditions, engine fuel, maintenance and repair tasks, tool usage, operator involvement, and spark plugs.

Operating Conditions

In order to verify that maintenance or repair tasks have been performed properly, or to determine if there is a problem with a small engine, the engine must be run at full or partial load after completing a test. When running an engine for testing or normal operation, the following safety practices must be observed:

- Only run the engine outdoors or in a well-ventilated area.
- Before operating equipment on land covered with dry grass or brush, install a spark arrestor. A *spark arrestor* is a component in the exhaust system of a small engine that redirects the flow of exhaust gases through a screen to trap sparks discharged from the engine. *See Figure 1-4.*

CARBON MONOXIDE (CO) METERS

- Digital display
- Handle

Fluke Corporation

Figure 1-3. A CO meter can be used to alert individuals in the area of the presence of CO indoors before it reaches lethal levels.

SPARK ARRESTORS

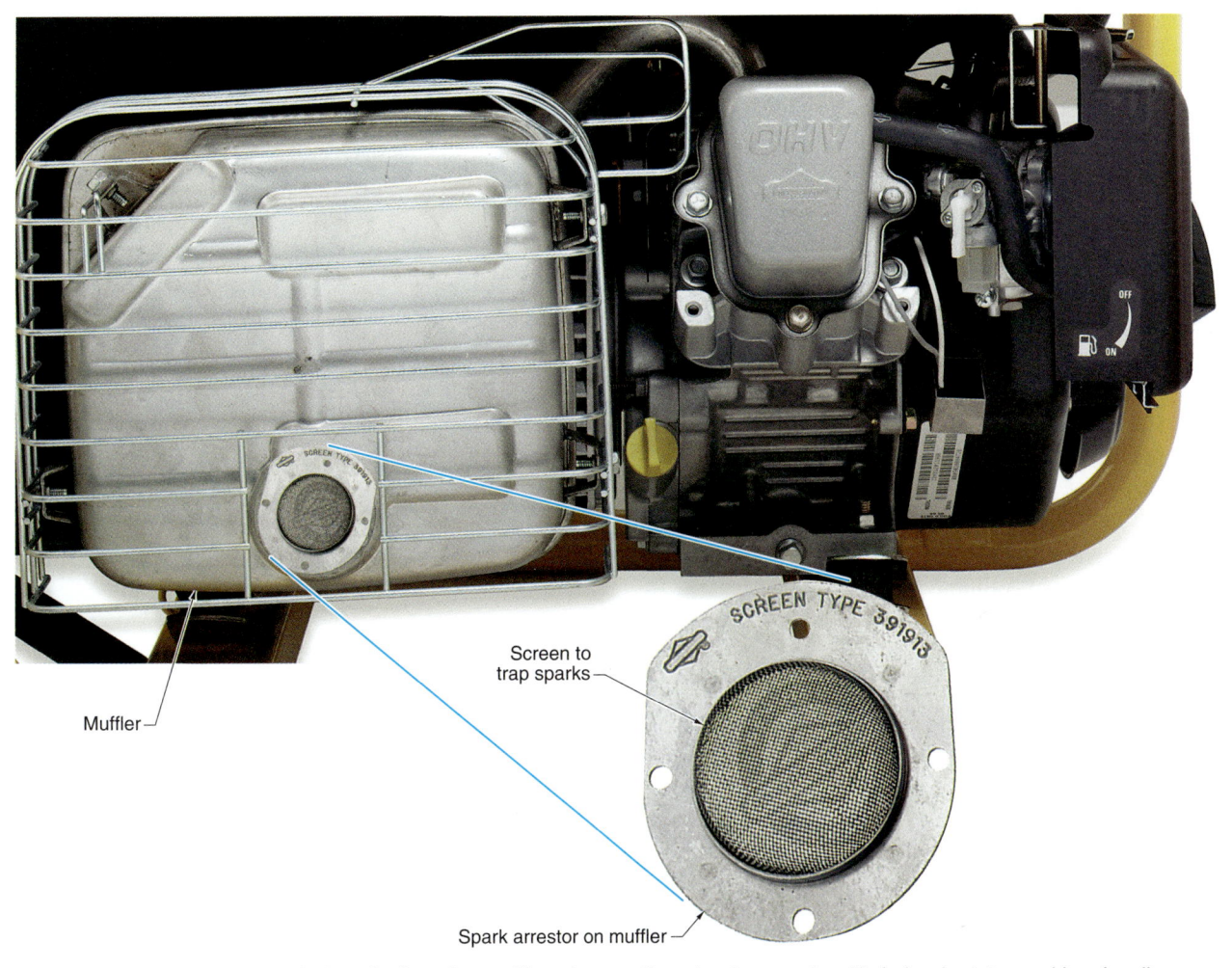

Figure 1-4. A spark arrestor is installed on the muffler of a small engine to comply with federal, state, and local ordinances regarding fire safety.

- Keep combustible and flammable materials a suitable distance from the muffler.
- Verify that the muffler is properly connected before starting the engine.
- Place equipment on a flat, level surface while it is in operation. Never tilt running equipment at a sharp angle.
- Do not operate the engine near gasoline or other combustible or flammable materials.
- Pull the starter cord slowly until resistance is felt, then pull rapidly to start. This helps to prevent injuries to hands and arms.
- Do not crank the engine with the spark plug removed. If the engine is flooded, place the throttle in the FAST position (sometimes identified by a rabbit symbol) and crank until the engine starts.
- Avoid running the engine at high speeds or in excess of the manufacturer's specifications.
- Keep hands and feet away from moving parts on the engine or equipment.
- To shut off the engine, gradually reduce engine speed. Turn the key to the OFF position, or move the controls to the OFF or STOP position.
- Shut off the engine before leaving the area. Never leave running power equipment unattended.

8 SMALL ENGINE AND EQUIPMENT MAINTENANCE

Engine Fuel Safety

The only place engine fuel and sparks should interact is in a combustion chamber. To reduce fire hazards, the following safety practices must be observed:

- Keep the engine and fuel tank a suitable distance from open flames, smoke, and combustible and flammable materials.
- Store fuel in Underwriter's Laboratory® (UL®)-or Canadian Standards Association (CSA)-approved, no-spill containers. Label the containers of flammable material for quick identification. *See Figure 1-5.*
- Avoid using spark-generating power tools and equipment where fuel vapors may be present.
- Allow the engine to cool before removing the fuel cap or filling the tank. Never fill a fuel tank while the equipment is in operation.
- Replace a fuel line or fitting if it leaks or is cracked.
- Keep fuels, solvents, and other flammables out of the reach of children.
- Store in a cool well ventilated area, never inside the home.

Proper Fuel Transportation. Spilled or dripping fuel can cause harm to individuals, equipment, and the environment. The occurrences of spills can be reduced by using UL®- or CSA-approved, no-spill fuel containers for transporting and storing fuel and other flammable liquids. Fuel containers are sometimes color-coded for the type of fuel they contain, such as red for gasoline, blue for kerosene, and yellow for diesel fuel.

Many steel fuel containers have built-in spark arrestors. An additional feature found on some fuel containers is a filter. A filter can help prevent debris and other contaminants from entering the engine fuel system.

FUEL STORAGE SAFETY

UL®- or CSA-approved, no-spill container

Fuel cannot flow through spout without operator assistance

No-Spill Fuel Containers

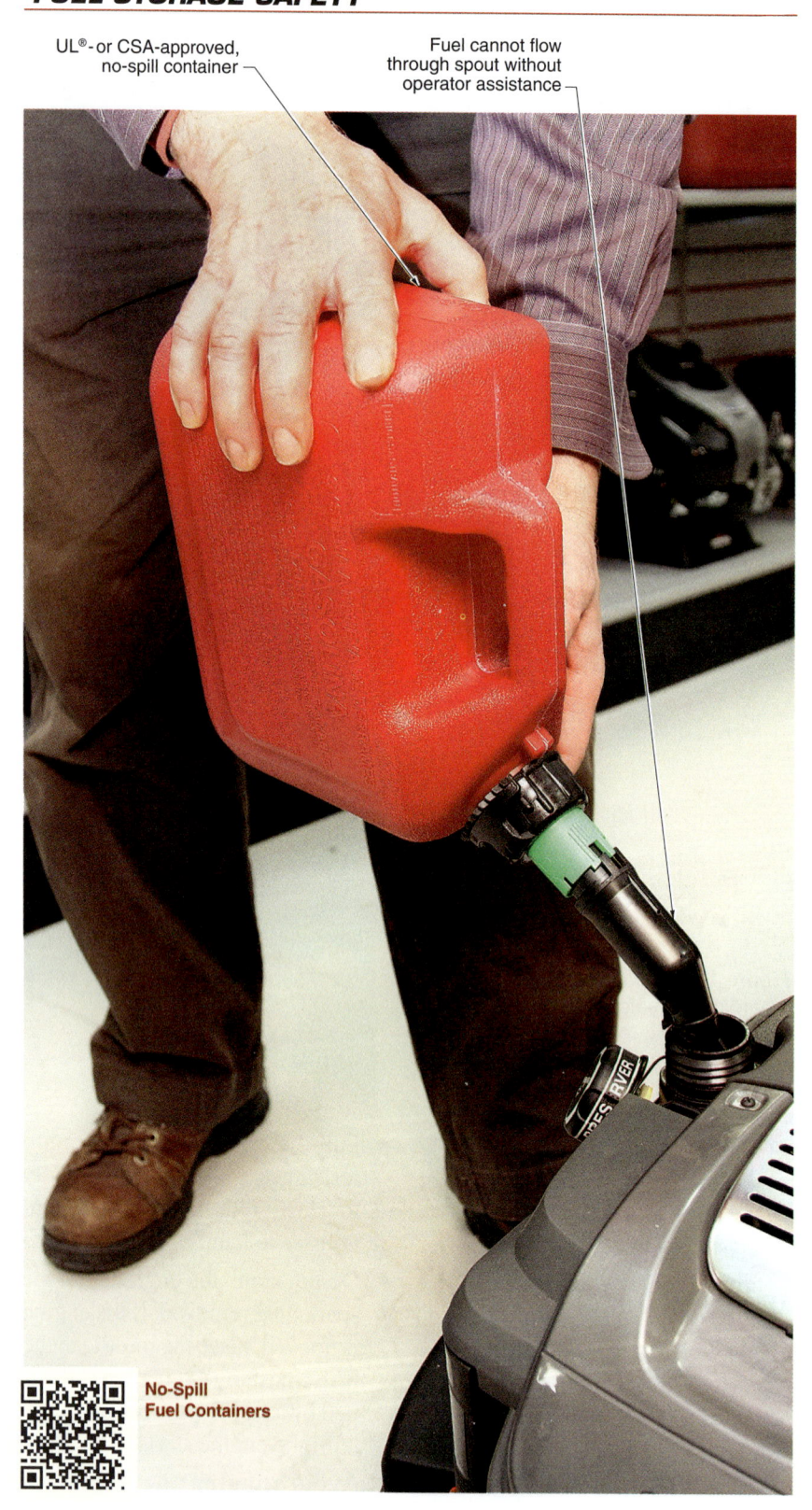

Figure 1-5. When working with equipment powered by small engines, proper safety practices require that fuel be stored only in UL®- or CSA-approved, no-spill containers.

When filling a fuel container or fuel tank of power equipment, the container or power equipment should be placed outdoors on a flat, level surface and away from appliances, heaters, and other sources of open flame or heat. A fuel container should never be filled while inside the trunk or on the truck bed of a vehicle.

During transport, fuel containers must be secured in an upright position and tightly sealed. A fuel container can be secured with a bungee cord inside the trunk or on the truck bed of a vehicle. **See *Figure 1-6*.** There should be ample ventilation to prevent fumes from the fuel from building up in the passenger compartment or trunk of the transport vehicle. Static electricity in these areas could ignite the fumes.

Maintenance and Repair Safety

Since small engines require flammable liquid to operate and can create a large amount of energy, many safety hazards are present. To safely perform small engine maintenance and repair tasks, the following safety practices must be observed:

- Verify that there is ample work space to perform the tasks required and maneuver around the engine, with easy access to tools and fire safety equipment.
- Have the correct tools for each job readily available.
- Keep an approved, fully charged fire extinguisher in an easily accessible location near the work area.

Fuel Transportation

- Be familiar with engine shutoff procedures for quick response in an emergency.
- If possible, disengage cutting blades, wheels, and other equipment before starting the engine.
- Disconnect the spark plug wire to prevent accidental starting while servicing the engine.
- Always disconnect any cable from the negative battery terminal when servicing an electric starter motor.
- Verify that the spark plug or spark tester is attached to the engine before cranking.
- Avoid contact with hot engine parts such as the muffler, cylinder heads, or cooling fins.
- Never strike the flywheel with a hammer or other hard object; such action may cause damage.
- Verify that the air cleaner assembly, blower housing, and muffler are properly connected before starting the engine.
- Remove any fuel from the tank and close the fuel shutoff valve (if equipped) before transporting the engine.
- Never operate an engine in an area where fuel has spilled.
- Use only original equipment manufacturer (OEM) replacement parts, as they are manufactured to exacting and EPA-compliant quality standards. Replacement parts installed on an engine that are not from the OEM may result in noncompliant operation.
- Maintain engine speed settings within manufacturer specifications. Operating the engine at speeds above the recommended maximum speed can damage or destroy the engine.

FUEL TRANSPORTATION

Cap tightly sealed

Fuel container in upright position

Bungee cord secured to truck bed

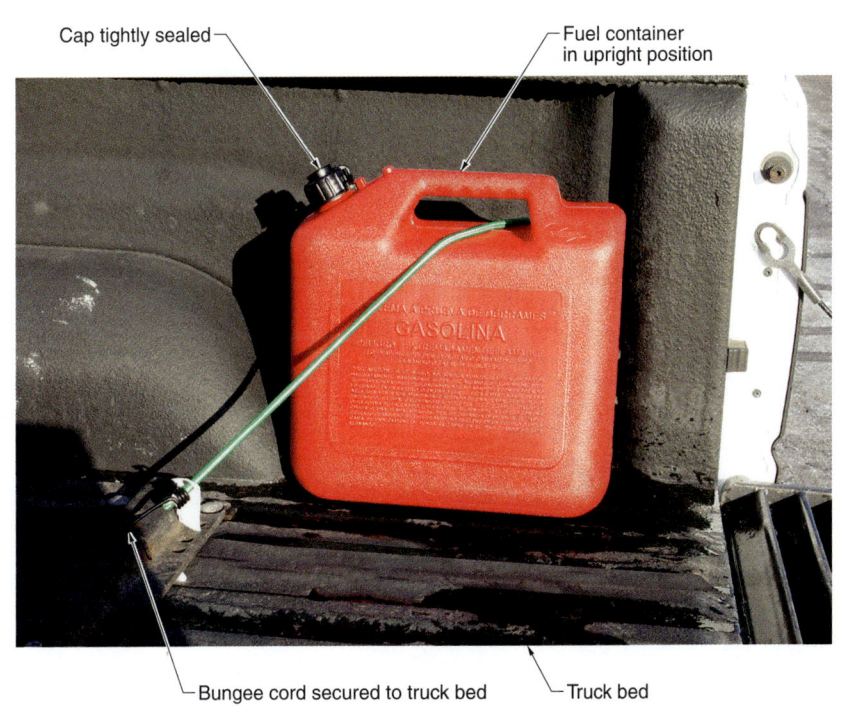

Truck bed

Figure 1-6. During transport, fuel containers must be secured in an upright position and tightly sealed.

Tool Safety

When working with tools, there is always the possibility of accidents occurring. Safety risks when using hand tools can be reduced by following hand tool safety rules as follows:

- Keep tools sharp and in proper working condition. Look for signs of wear that could cause an injury, such as a pitted hammer face, damaged insulation, or splintered handles.
- Point cutting tools away from the body during use.
- Organize tools to protect and conceal cutting edges.
- Never use a hammer on another hammer. The impact of the hardened surfaces may cause the heads to shatter.
- Do not carry tools in pockets. Transport sharp tools in holders or with the blades pointed down.
- When removing fasteners, pull the tool toward the body or push the tool away from the face.

Safety risks when using power tools, such as a portable drill, can be reduced by following basic power tool safety rules. The rules for power tool safety include the following:

- Follow all manufacturer-recommended operating instructions.
- Use UL®- or CSA-approved power tools that are manufactured in compliance with NEC® requirements.
- Do not use electrical tools in or near wet or damp areas.
- Use power tools that are double-insulated or have a third conductor grounding terminal to provide a path for fault current.
- Ensure the power switch is in the OFF position before connecting to a power source.
- Arrange cords and hoses to prevent accidental tripping.
- Stand clear of power tools in operation. Keep hands and arms away from moving parts.
- Do not use any tool near flammable materials such as gasoline.

TECH FACT

When working on a small engine and/or related equipment, loose-fitted clothing and long hair must be secured to prevent getting caught on rotating parts. Jewelry such as wrist watches, necklaces, and rings should be removed. Watches and rings are excellent conductors of electricity and may cause serious burns if contact is made with an electrical circuit. Soiled protective clothing should be washed to reduce flammability hazards.

Tool Safety

Operator Safety

Fire, electrical shock, and asphyxiation are not the only results of unsafe practices when performing work on small engines. Injuries to the eyes, ears, lungs, hands, feet, arms, legs, and back can also be caused by accidents from unsafe work practices. These types of injuries can be prevented by wearing proper personal protective equipment (PPE). *See Figure 1-7.* Safety rules that prevent injuries include the following:

- Keep hands, feet, and clothing away from moving engine and equipment components.
- Use eye protection, such as goggles or safety glasses with side shields, when working with engines or power tools.
- Wear ear protection to reduce the risk of gradual hearing loss from exposure to engine noise.
- Wear a face shield with saftey glasses, when working with chemicals.
- Wear neoprene gloves to protect against harmful chemicals.
- Wear heavy-duty leather work gloves to protect against heat and sharp objects.
- Wear steel-toe safety shoes to protect against falling objects. Safety shoes also have soles that are impervious to fuel, grease, and oil.
- Use proper lifting techniques when moving heavy or cumbersome objects.

PERSONAL PROTECTIVE EQUIPMENT (PPE)

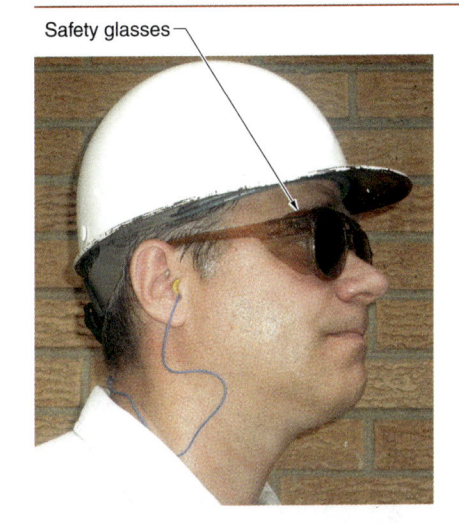

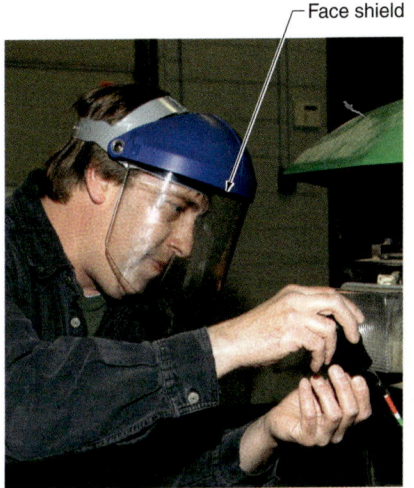

Figure 1-7. Injuries to the eyes, ears, lungs, hands, feet, arms, legs, and back can be prevented by wearing proper PPE.

Spark Plug Safety

A small engine cannot start or operate without several different components. The easiest method that can be used to prevent accidental starting of the engine is to disable the spark plug. For complete protection from the accidental starting of an engine, the spark plug lead (wire) should be disconnected and secured away from the spark plug when performing maintenance or repair tasks. *See Figure 1-8.*

Chapter 1 Quick Quiz®

SPARK PLUG SAFETY

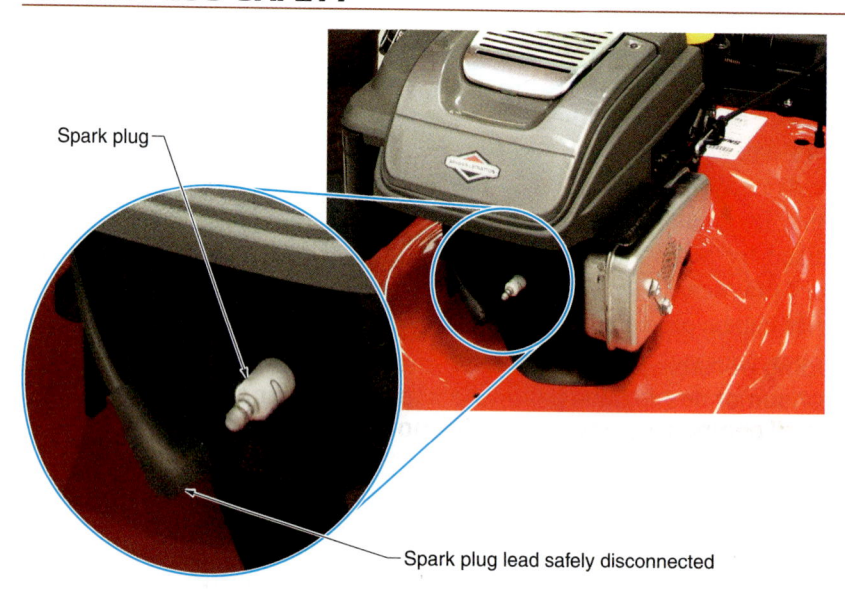

Figure 1-8. For protection from accidental starting of an engine during maintenance, the spark plug lead should be disconnected away from the spark plug.

History of Small Engines and Related Applications

2

INTRODUCTION

Small engines have improved the quality of life by providing portable power when and where it is needed. As consumer needs have grown, so has small engine use. Small engines are used to provide power for lawn mowers, snow throwers, edgers, rototillers, string trimmers, leaf blowers, and other equipment that make outdoor yard work easier to perform. They also provide power to equipment used for house painting, power washing, masonry installation, and a variety of other outdoor home projects. Demand for these types of machines increases each year.

HISTORY OF OUTDOOR POWER EQUIPMENT

Outdoor power equipment (OPE) has been in use in one form or another for about 200 years. Engines early in this time period were powered by steam, diesel fuel, or gasoline. However, the production of American-built automobiles in the early 1900s made gasoline-powered engines more desirable than steam- and diesel-powered engines.

Although automobiles were too expensive for most Americans to own, the idea that gasoline-powered engines could simplify and improve the daily quality of life was popularized. Hoping to capitalize on this idea, many manufacturers began to develop smaller, less expensive equipment that could be powered by gasoline engines. Many of these manufacturers, including Briggs & Stratton, developed engines that could be used on outdoor power equipment such as lawn mowers, fruit gathering machines, and garden cultivators. *See Figure 2-1.*

Small gasoline engines and outdoor power equipment were used earliest in the agriculture industry. Harvesting equipment powered by small gasoline engines was designed to reduce manual labor through the increased efficiency of harvesting crops in agricultural fields.

Briggs & Stratton History

13

14 SMALL ENGINE AND EQUIPMENT MAINTENANCE

EARLY SMALL ENGINE APPLICATIONS

FRUIT GATHERING MACHINES

LAWN MOWERS

GARDEN CULTIVATORS

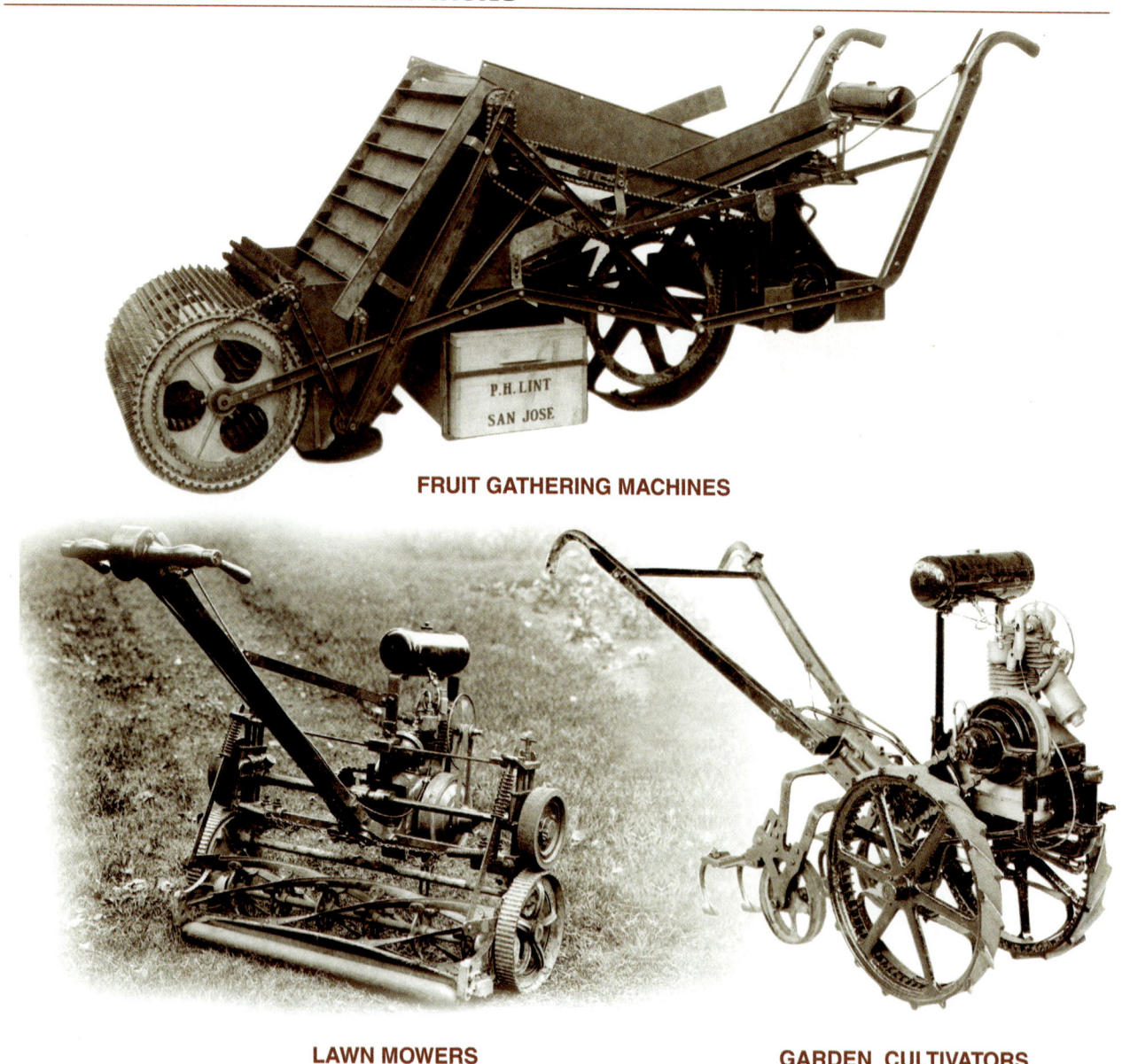

Figure 2-1. Early small engine applications included fruit gathering machines, lawn mowers, and garden cultivators.

TECH FACT
One of the most successful early applications of the Briggs & Stratton engine was the motor wheel. Motor wheels were single wheels attached to small engines. They were used to provide motorized power to devices such as bicycles, go-carts, scooters, and snow sleds. Motor wheels were most commonly used on bicycles.

In the early 1900s, small cast iron gasoline engines manufactured primarily by Briggs & Stratton dominated the market. They were used to provide power to cultivating machines such as the small tractors and rototillers used on farms. Various types of outdoor power equipment, such as water pumps for irrigation, sprayers for pest control, and engines for power generation, were commonly used on farms both large and modest in size. Briggs & Stratton developed several designs of small engines. The design of the Model P engine was the earliest and most commonly used design. ***See Figure 2-2.***

EARLY SMALL ENGINE TYPES

MODEL K

MODEL H

MODEL P

MODEL FG

Figure 2-2. Briggs & Stratton developed several designs of small engines, with the design of the Model P engine being the earliest and most common design.

16 SMALL ENGINE AND EQUIPMENT MAINTENANCE

The Model P engine and other similar engines used in farming applications provided portable power to equipment that was normally horse drawn such as cultivators, hay balers, and threshers. However, not all equipment and appliances were used for agricultural applications. For example, one of the earliest residential applications of the small engine was the gasoline engine-powered clothes washing machine.

Early Consumer Equipment

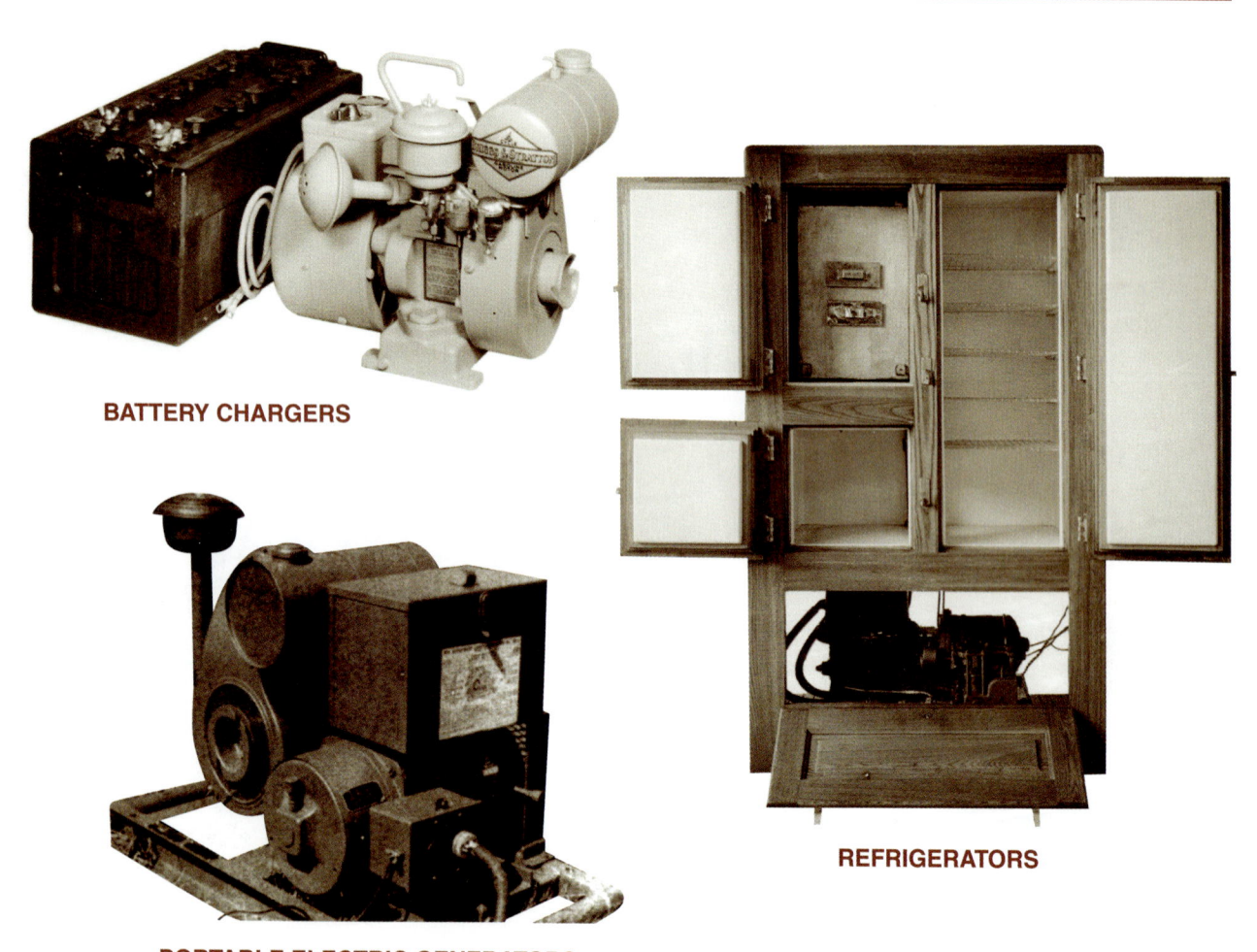

It was the simplicity of design and practical nature of the equipment that launched the modern outdoor power equipment industry and it began with the clothes washing machine. During the postwar era of the 1920s and 1930s, the use of outdoor power equipment for agricultural and residential applications grew at an astonishing rate. *See Figure 2-3.* In addition to clothes washing machines and push reel lawn mowers, equipment that was commonly powered by small engines during this time period included the following:

- water pumps
- refrigerators
- cream separators
- milking machines
- portable electric generators
- battery chargers
- portable and isolated electric light plants
- railroad rail spreaders
- wood cutting saws

EARLY CONSUMER EQUIPMENT

BATTERY CHARGERS

PORTABLE ELECTRIC GENERATORS

REFRIGERATORS

Figure 2-3. Early consumer equipment powered by small engines included the portable battery charger, refrigerator, and portable electric generator.

Clothes Washing Machines. It was the popularity and practicality of the clothes washing machine powered by small engines that opened the market for the consumer use of outdoor power equipment. Powering a clothes washing machine, which was typically used outdoors, with a ½ HP, 1750 rpm air-cooled gasoline engine was an innovation that significantly reduced the effort and time required to wash clothes.

The engine was mounted on a carriage assembly located beneath the tub and wringer assembly. With some models, an exhaust tube that could be routed outdoors allowed the washing machine to be used indoors. Several designs of the engine-powered clothes washing machine were available. *See Figure 2-4.*

SMALL-ENGINE-POWERED CLOTHES WASHING MACHINES

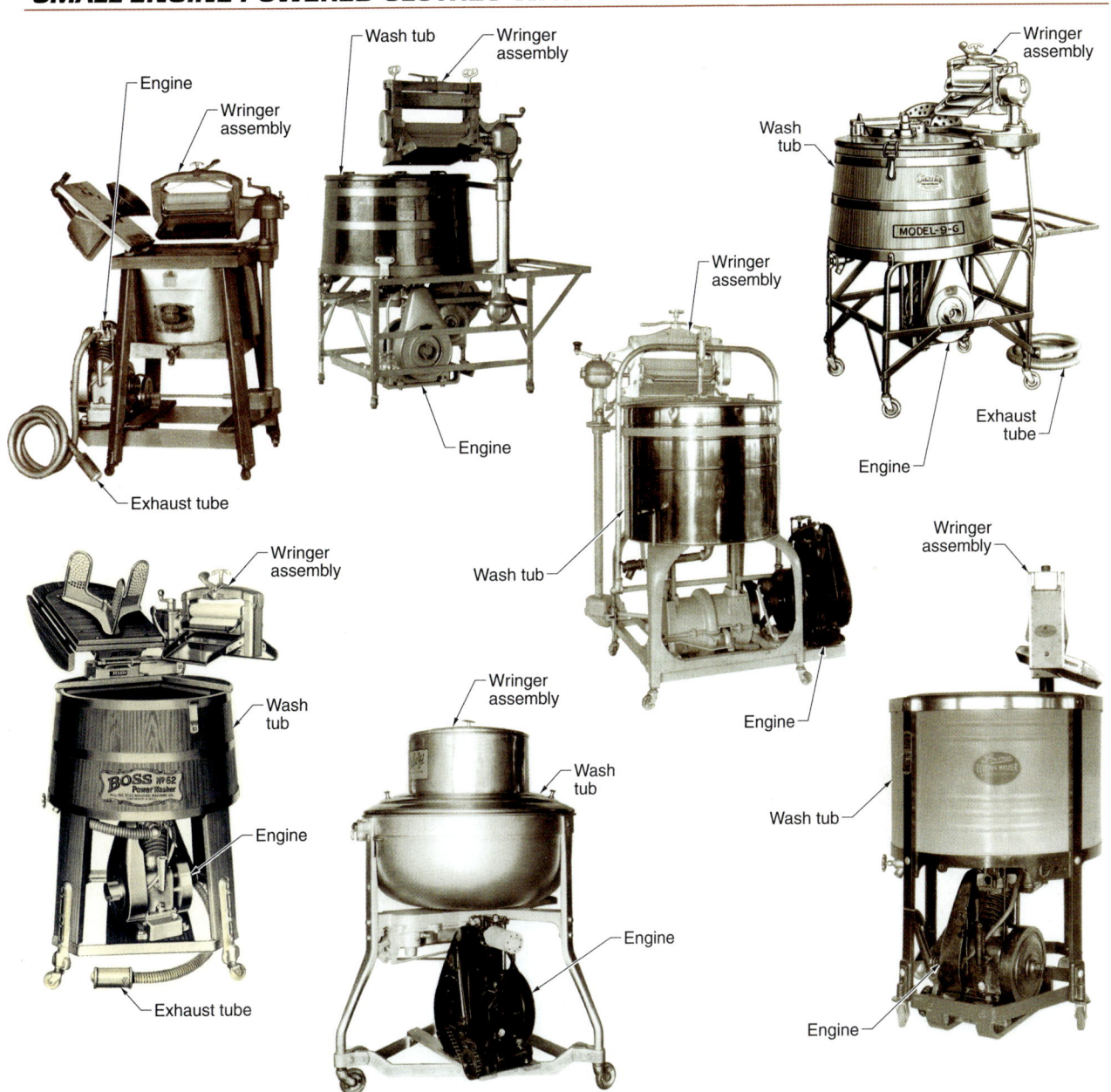

Figure 2-4. The popularity of small-engine-powered clothes washing machines opened the market for consumer use of outdoor power equipment.

Push Reel Lawn Mowers. As usage of equipment powered by small engines increased, many other applications were developed for outdoor power equipment, which improved the quality of life. Postwar population growth and housing construction throughout the country initiated the need for equipment that could be used for residential maintenance projects, such as lawn care.

A *push reel lawn mower* is a lawn cutting device with multiple cutting blades attached to a rotating central shaft between two wheels. The blades are connected together as a horizontal cylinder. As the wheels move, the shaft between them rotates, allowing the blades to rotate. As the blades rotate, they trap grass between the sharp blades and a stationary knife at the base of the reel. As the multiple blades pass the stationary knife, the grass caught between the two is sheared off at the desired length.

The push reel lawn mower was originally pulled by a horse or pushed manually by the operator. However, the equipment gained popularity when a small air-cooled gasoline engine was added to the design to reduce operator fatigue and increase lawn cutting production. ***See Figure 2-5.***

TECH FACT
Due to changing the composition of engines from cast iron to aluminum, the versatility of small engines increased. By 1967, the Briggs & Stratton Company had sold over 50 million small engines.

PUSH REEL LAWN MOWERS

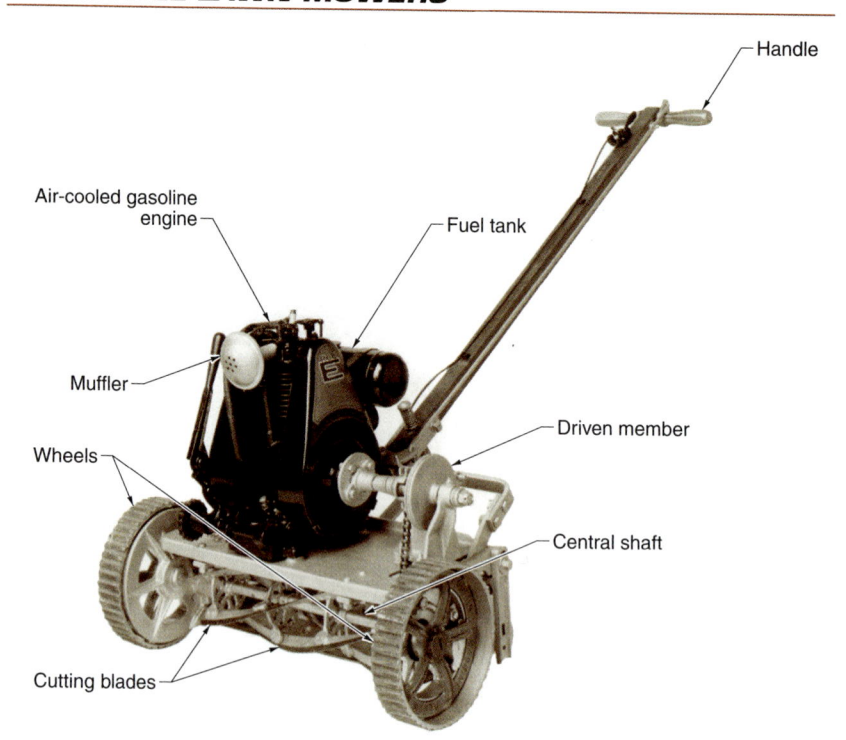

Figure 2-5. A small air-cooled gasoline engine was added to the original design of the push reel lawn mower to reduce operator fatigue and increase lawn cutting production.

Consumer Market Growth

While cast iron small engines were effective for stationary equipment, they were heavy and impractical for mobile equipment such as lawn mowers. It was not until 1953 that the first die-cast aluminum alloy engine was introduced, allowing engine-powered lawn mowers and other equipment to be lighter and easier to operate. The introduction of aluminum alloy engines made cast iron engines virtually obsolete. Because of this development, the sales growth of engines used for mobile outdoor power equipment increased significantly.

During the 1950s, increasing growth in the residential market and the popularity of a manicured lawn led to a large and growing outdoor power equipment market, which increased demand for more consumer friendly equipment designs and reliability. As the number of residential lawns and gardens increased, the outdoor power equipment industry continued to deliver more consumer-focused engine-powered machinery to keep pace with the growing market.

Two of the most significant pieces of equipment introduced during this time period were the rotary lawn mower and the lawn and garden tractor. The popularity of these devices led to the development of additional outdoor power equipment for consumer use such as string trimmers, leaf blowers, snow throwers, and pressure washers. ***See Figure 2-6.***

COMMON OUTDOOR POWER EQUIPMENT (OPE)...

STRING TRIMMERS

LEAF BLOWERS

SNOW THROWERS

PRESSURE WASHERS

Figure 2-6. The demand for lawn mowers and lawn and garden tractors led to the development of additional OPE such as string trimmers, leaf blowers, snow throwers, and pressure washers.

20 SMALL ENGINE AND EQUIPMENT MAINTENANCE

. . . COMMON OUTDOOR POWER EQUIPMENT (OPE)

ROTOTILLERS

LAWN EDGERS

WOOD CHIPPERS

PORTABLE GENERATORS

Figure 2-6. (Continued)

Rotary Lawn Mowers. A *rotary lawn mower* is a grass cutting device that uses one or more flat, horizontal blades to cut grass with a high-speed circular rotation. The leading edge of the blade is sharpened at an angle to perform the cutting action. Most of the grass cutting implements or attachments became equipped with single or multiple rotating blades. This design change, along with the addition of a lightweight aluminum engine, made lawn care easier and less time consuming. This design is currently used on all types of modern lawn mowers. Early Briggs & Stratton aluminum engines were produced in the 1950s and mainly used to reduce the weight of rotary lawn mower engines. *See Figure 2-7.*

TECH FACT

After the rotary lawn mower was developed, the most significant development for Briggs & Stratton was the aluminum engine. Although a lawn mower equipped with a cast iron engine could cut grass as efficiently as one with an aluminum engine, the weight of the cast iron engine (approximately 40 lb) made operation of the lawn mower difficult. The aluminum engine was developed to reduce the total weight of the equipment and withstand the demands of lawn mowing.

EARLY ALUMINUM ENGINES

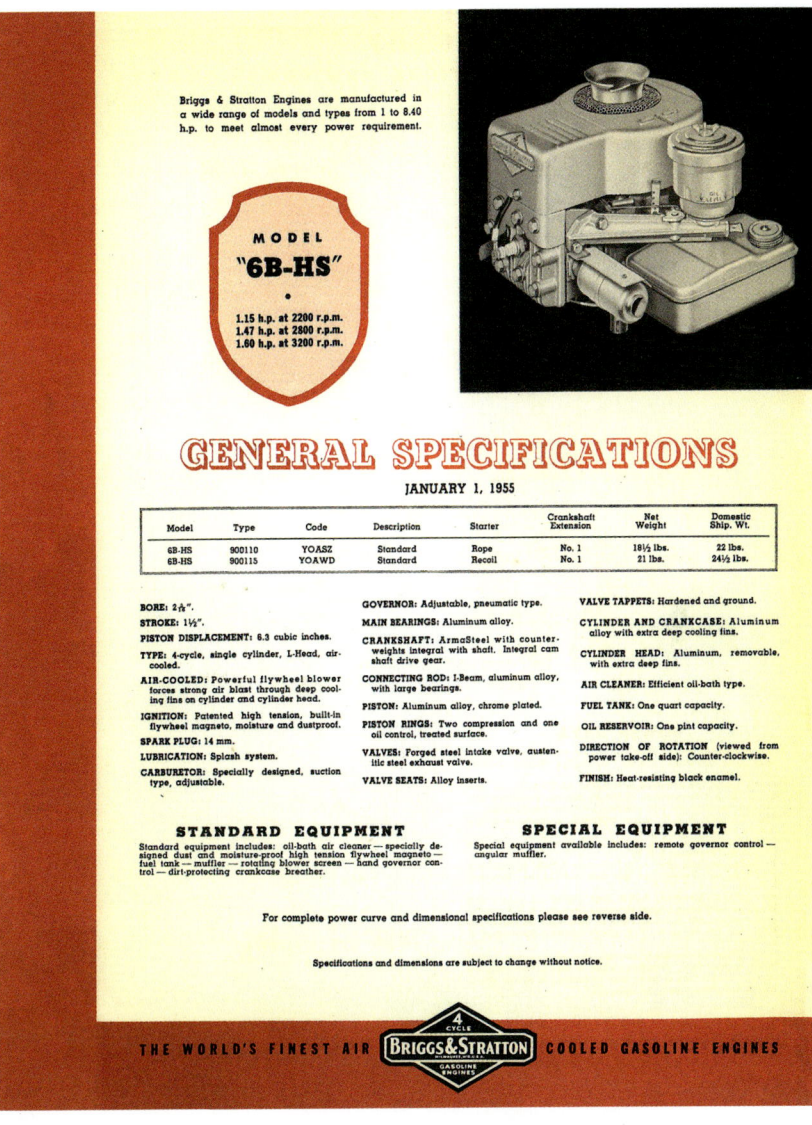

Figure 2-7. Early Briggs & Stratton aluminum engines were produced in the 1950s and mainly used to reduce the weight of rotary lawn mower engines.

Lawn and Garden Tractors. A *lawn and garden tractor* is a gasoline-powered four wheel machine that allows the operator to ride on the equipment as it performs work. A lawn and garden tractor typically has a rotary mower deck attached to its undercarriage that contains one to three horizontal cutting blades. The cutting blades are evenly spaced across the width of the mower deck.

A lawn and garden tractor is typically powered by an 8 HP to 31 HP gasoline engine, which provides sufficient power to propel the tractor across the lawn while the blades cut the grass. Various implements may be connected to lawn and garden tractors to perform specific types of work. For example, many lawn and garden tractors can have a snow thrower, plow blade, front end loader, or cultivator attachment (powered by the engine) connected to them. Designed primarily for acreage and large lawns, lawn and garden tractors make up about 25% of the lawn mower market in North America. *See Figure 2-8.*

LAWN AND GARDEN TRACTORS

Figure 2-8. Lawn and garden tractors were originally developed to mow acreage and large lawns.

Chapter 2
Quick Quiz®

Outdoor Power Equipment Applications

3

INTRODUCTION

Outdoor power equipment (OPE) is most commonly used for lawn and garden maintenance applications. The types of equipment used for lawn and garden maintenance applications include ride-on equipment, such as lawn and garden tractors, handheld equipment, such as leaf blowers and string trimmers, and portable equipment that must be transported to the location of use, such as portable generators and wood chippers.

Basic preventative maintenance is required for equipment to remain in good operational condition. Because preventative maintenance involves the regular and proper cleaning, adjustment, and repair of equipment, maintenance quality is directly related to the longevity of any equipment.

OUTDOOR POWER EQUIPMENT APPLICATIONS

In modern terminology, the term "small engine" applies to more than walk-behind lawn mowers. For example, four-stroke cycle small engines range from 2.4 HP engines for light-duty equipment, such as pressure washers, to 31 HP engines for tractors and other heavy-duty equipment. *See Figure 3-1.* A four-stroke cycle small engine is the most common type of small engine used on outdoor power equipment (OPE). Nearly every consumer household owns or uses a four-stroke cycle small engine on some type of OPE.

Advances in lawn and garden equipment have changed the way many homeowners care for their homes. Owning a home with a large lawn no longer requires hiring a landscaper. Comfortable, powerful, easy-to-use lawn equipment makes it possible for homeowners to care for their lawns themselves. The cost of the equipment can be recovered in just a few years.

Due to the affordability and simple design of Briggs & Stratton small engines for equipment, consumers can complete outdoor projects that once were too difficult and time-consuming to perform. Small engines are currently used to provide power for rototillers, wood chippers, garden tractors, compressors, standby generators, and other products that an increasing number of residential and commercial users operate. Generally, OPE is categorized as lawn and garden, consumer, and industrial and commercial equipment.

Small Engine Oil Selection

FOUR-STROKE CYCLE SMALL ENGINES

2.4 HP

31 HP

Figure 3-1. Four-stroke cycle small engines range from 2.4 HP to 31 HP.

Lawn and Garden Power Equipment

Lawn and garden power equipment is power equipment that is used exclusively outdoors in residential or light commercial environments. Examples of lawn and garden equipment include rotary lawn mowers, zero-turn (ride-on) lawn mowers, and lawn and garden tractors. This equipment is used mainly for lawn maintenance such as grass cutting.

Walk-Behind Rotary Lawn Mowers. The most common type of powered lawn and garden equipment is a walk-behind rotary lawn mower. Walk-behind rotary lawn mowers may be self-propelled or standard push mowers. There are many different designs of walk-behind rotary lawn mowers, but each consists of an engine, handle, wheel, mower deck, and blade. *See Figure 3-2.*

Lawn Mowers

TECH FACT

Engine output is measured based on torque and determined by calculating horsepower. Torque is measured in pound-feet (lb-ft) or newton-meters (Nm). Torque is the measure of the ability of an engine to do work. Power is a calculation of the rate at which work is done. For example, the amount of torque determines if an engine can drive a lawn mower through tall grass. The amount of power determines how quickly the mower progresses through the grass.

WALK-BEHIND ROTARY LAWN MOWER PARTS

Figure 3-2. The basic parts of a walk-behind rotary lawn mower include an engine, handle, operator presence bail, wheels, mower deck, and blade.

26 SMALL ENGINE AND EQUIPMENT MAINTENANCE

Standard push mowers are propelled by the operator and cut grass using a single rotating blade, which is sharpened on the leading edge. Blade rotation speed is controlled by the operator with an engine speed control device. Various models of push-type walk-behind rotary lawn mowers are designed specifically for residential or commercial use. *See Figure 3-3.*

TECH FACT
The blade speed on a walk-behind rotary lawn mower is typically set to the manufacturer's suggested top no-load speed. The engine is to be run at this speed. Ground speed is controlled by the operator by either slowing walking speed or using the self-propelled speed control.

WALK-BEHIND ROTARY LAWN MOWERS

RESIDENTIAL TYPE

COMMERCIAL TYPE

Figure 3-3. Walk-behind rotary lawn mowers are designed specifically for residential or commercial use.

Remote Speed Control

Another common design of a walk-behind rotary lawn mower is a self-propelled rotary lawn mower. This lawn mower uses a drive system to propel the equipment across a lawn, reducing operator effort. The grass cutting action of the self-propelled lawn mower and the push lawn mower is the same. Also, the elements are the same, with the exception of the self-propelled drive system. A self-propelled drive system is commonly equipped with a V belt or friction disc drive system. By using a speed control device connected directly to the engine, the operator can select a comfortable mowing speed appropriate for the type of terrain. *See Figure 3-4.*

Zero-Turn Lawn Mowers. In recent years, the market growth of zero-turn lawn mowers has increased due to the increase of homeowners with large lawns to maintain. A *zero-turn lawn mower* is a ride-on lawn mower with a tight turning radius that allows the operator to control forward and backward motion using control handles. It turns in a 360° radius without the steering limitations of lawn and garden tractors. A zero-turn lawn mower consists of a chassis, rear-mount engine, centrally located operator seat, and mower deck. Also included in the design are two control arms for steering and speed control and front wheels designed to turn 360°. *See Figure 3-5.*

An advantage of zero-turn lawn mowers is that they can cut closely around obstacles, pivoting 180°, without leaving circles of uncut grass. A disadvantage of zero-turn lawn mowers is that implements and attachments are very limited.

SPEED CONTROL DEVICES

Figure 3-4. A speed control device is used by a lawn mower operator to select a comfortable mowing speed appropriate for the type of terrain.

ZERO-TURN LAWN MOWERS

Figure 3-5. Zero-turn lawn mowers are designed to cut closely around obstacles, pivoting 180°, without leaving circles of uncut grass.

28 SMALL ENGINE AND EQUIPMENT MAINTENANCE

Typically, each rear wheel of a zero-turn lawn mower is connected to a hydrostatic transmission. *See Figure 3-6.* The hydrostatic transmission controls wheel rotation by using pressurized transmission fluid. A steering and speed control lever is used to control the flow of transmission fluid and the rotational speed (forward/reverse direction) of each drive wheel. Maximum lever movement results in maximum fluid flow and a rapidly turning wheel.

The direction of the lawn mower is determined by the speed and direction of the drive wheels. If one wheel stops while the other turns, or if the wheels turn in opposite directions, the mower pivots. If one drive wheel turns more rapidly than the other, the lawn mower travels along a curved path. If both wheels turn at the same speed, the lawn mower travels along a straight path.

Lawn and Garden Tractors. Lawn and garden tractors are commonly used to cut large areas of grass while riding the equipment. They are designed for the addition of attachments and implements, which can be added and removed by the operator to perform various outdoor tasks such as lawn sweeping, snow removal, and plowing. This capability makes them versatile enough to be used during any season. *See Figure 3-7.*

A lawn and garden tractor includes a chassis, mower deck, steering wheel, and operator seat. The main differences between lawn and garden tractors and zero-turn lawn mowers are the turning radius of the steering mechanisms and the variety of work that can be performed.

ZERO-TURN LAWN MOWERS—HYDROSTATIC TRANSMISSIONS

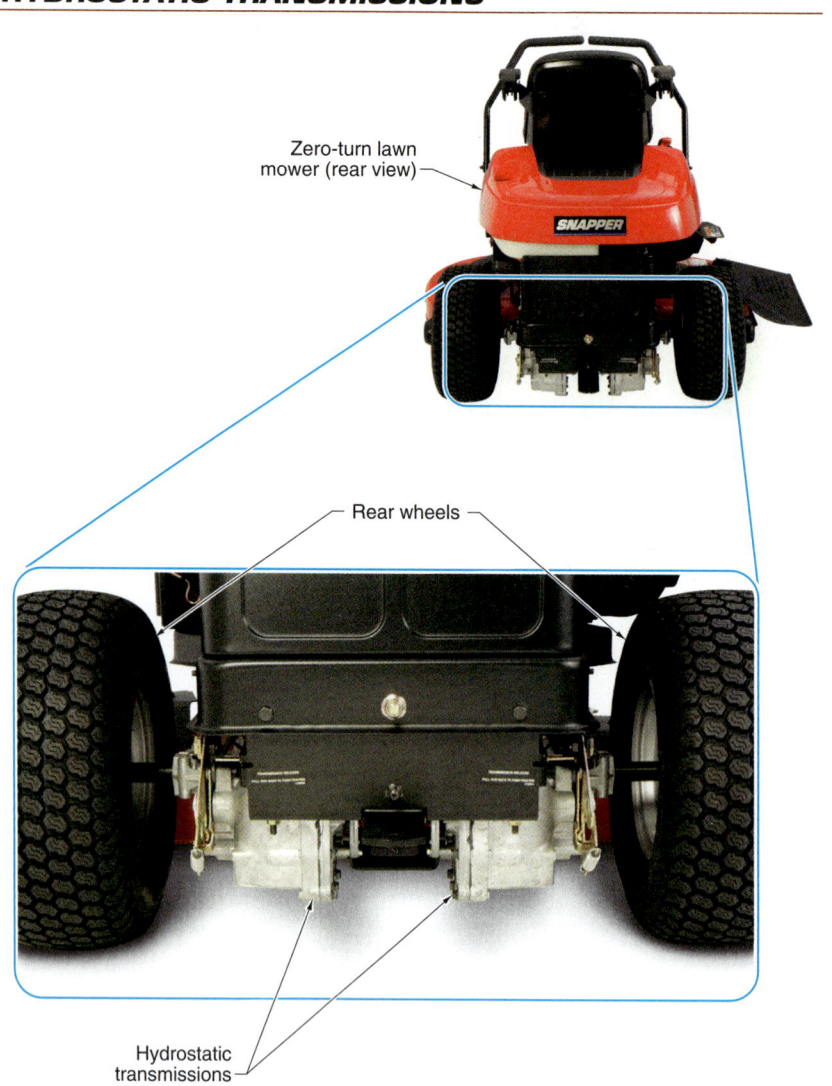

Figure 3-6. Each rear wheel of a zero-turn lawn mower is connected to a hydrostatic transmission.

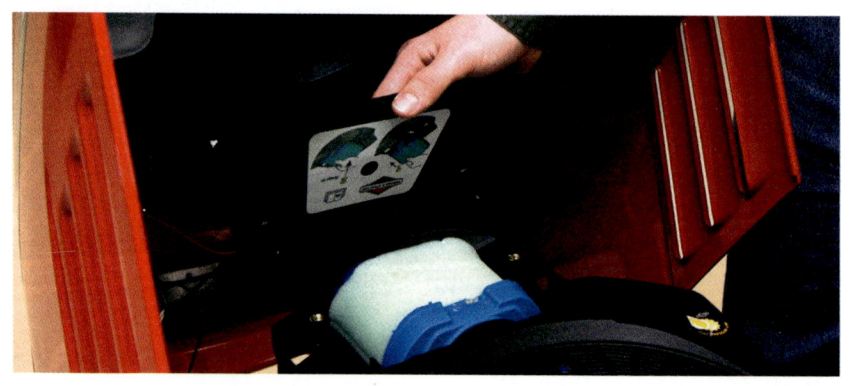

Engine air filters on lawn and garden tractors must be frequently inspected for cleanliness.

LAWN AND GARDEN TRACTORS — SEASONAL USES

MOWING LAWN (SPRING/SUMMER)

CLEARING SNOW (FALL/WINTER)

Figure 3-7. Because of the ability to accept different attachments and implements, lawn and garden tractors are versatile enough to be used during any season.

Consumer Outdoor Power Equipment

Consumer outoor power equipment is designed for nonprofessional and residential use. Common types of consumer equipment, not including lawn mowers, that use four-stroke cycle small engines include pressure washers, wood chippers, leaf blowers, string trimmers, and snow throwers. These devices are used primarily by homeowners to remove snow and debris, such as leaves, twigs, and branches, from lawns and walkways.

Pressure Washers. A *pressure washer* is a portable device that cleans surfaces with pressurized water. Many pressure washers are powered by air-cooled four-stroke cycle engines. They can provide pressures ranging from 1800 psi to 4200 psi.

A pressure washer consists of a four-stroke cycle engine, high-pressure pump, high-pressure hose, control wand, and nozzle. A pressure washer receives water from a supply through a garden hose and increases the water pressure using an engine-powered pump. The high-pressure water exits the pump through a hose and into a control wand, which is held by the operator. When the operator depresses the trigger on the wand, water exits through a small, adjustable orifice in the nozzle tip. The amount of water pressure created depends on the engine and pump size. *See Figure 3-8.*

Wood Chippers. A *wood chipper* is a device used to chop and shred organic materials into ¼″ particles. Many wood chippers are designed as free-standing machines commonly attached to wheels, which allow easy maneuvering. A typical wood chipper consists of a small four-stroke cycle engine, cutting disc with sharpened steel knives, debris intake chute, and debris discharge chute. Organic materials, such as twigs and branches several inches in diameter, are loaded into the intake chute by hand and shredded in the cutter system. The shredded material is commonly used as landscaping mulch. *See Figure 3-9.*

PRESSURE WASHERS

High-pressure water discharge — Nozzle with adjustable orifice — Control wand — Hand cart

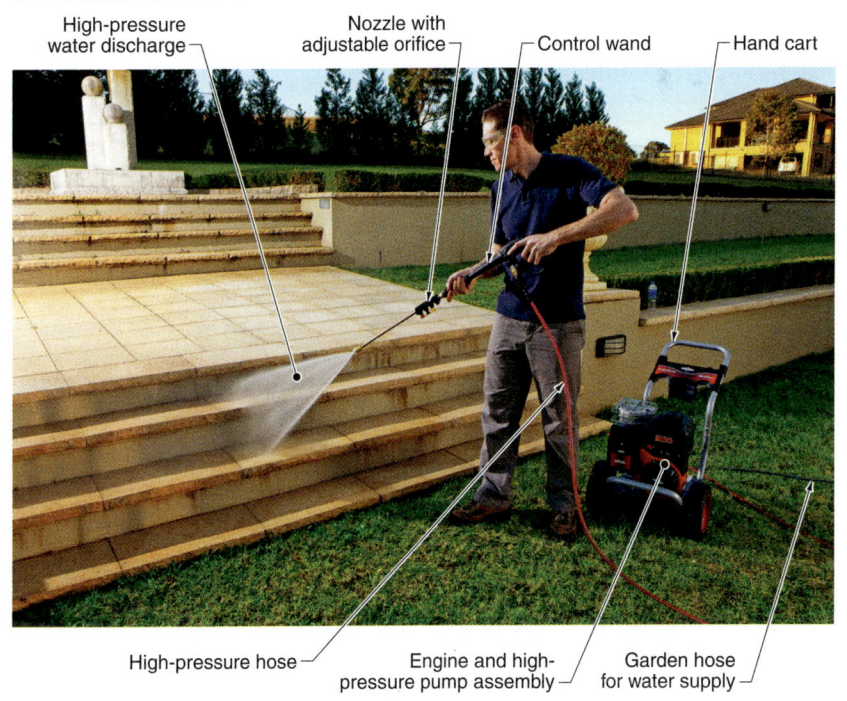

High-pressure hose — Engine and high-pressure pump assembly — Garden hose for water supply

Figure 3-8. Pressure washers provide pressurized water for the purpose of cleaning surfaces.

30 SMALL ENGINE AND EQUIPMENT MAINTENANCE

TECH FACT

The proper personal protective equipment (PPE), such as safety glasses or goggles and leather gloves, should always be worn when operating a wood chipper. Clothing worn should not be loose or frayed. Objects that could damage the wood chipper should never be placed in the intake or discharge chutes. Manufacturer's instructions should always be followed and hands and feet should be kept away from the intake and discharge chutes.

WOOD CHIPPERS

Figure 3-9. Wood chippers are used to chop and shred organic growth materials into small particles.

Labels: Intake chutes; Four-stroke cycle engine; Wheels to allow mobility; Discharge chute with bag

Engines

Leaf Blowers. A *leaf blower* is a motorized device that creates high-speed and high-volume blowing air to move light debris away from driveways, walkways, roofs, and lawns. Leaf blowers are specialized debris management devices designed for consumer and commercial markets. Depending on the engine size, they can provide blowing air of 135 mph to 220 mph. Leaf blowers are available as walk-behind or handheld models. A leaf blower produces highly accelerated air discharged through a delivery chute. Most handheld leaf blowers powered by gasoline engines use two-stroke cycle engines, although many walk-behind models with four-stroke cycle engines are available. **See Figure 3-10.**

LEAF BLOWERS

Labels: Air-discharge delivery chute; Engine

Figure 3-10. Leaf blowers create high-speed blowing air to move light debris, such as grass clippings, leaves, and twigs, away from driveways, walkways, roofs, and lawns.

String Trimmers. A *string trimmer* is a powered handheld device that uses a flexible monofilament line for cutting grass and other plants near objects. A string trimmer is also referred to as a weed whacker or Weed Eater®. It consists of a cutting head at the end of a long shaft with a handle and sometimes a shoulder strap. *See Figure 3-11.* String trimmers may be powered by an electric motor or by a two- or four-stroke cycle engine. The engine of an engine-powered string trimmer is mounted on the opposite end of the shaft from the cutting head. An electric string trimmer typically has an electric motor in the cutting head.

An advantage of engine-powered string trimmers over electric motor-powered types is that all engine-powered string trimmers are cordless, while electric models must use either an extension cord or a rechargeable battery. Commercial-grade string trimmers, used for cutting roadside grass in large areas, can be heavy and must be suspended from the body by a harness. Both hands may be required to control the device. Commercial-grade string trimmers are often referred to as brush cutters.

Snow Throwers. A *snow thrower* is a machine that removes snow, as from a sidewalk or driveway, by using a rotating spiral blade to pick up the snow and propel it aside. Snow removal and management are common activities in colder climates. The two designs for snow throwers are single-stage and two-stage, which are similar in appearance.

STRING TRIMMERS

Figure 3-11. String trimmers have a cutting head with a flexible monofilament line for cutting grass and other plants near objects.

SNOW THROWERS

A single-stage snow thrower commonly uses a single rotating auger (disc). The auger of a single-stage snow thrower typically resembles a single curved blade. The auger sweeps snow into a snow collection hopper and discharges the snow from a chute at the top of the machine. The chute can be rotated to discharge the snow to an area away from the area that is being cleaned.

A two-stage snow thrower incorporates the single-stage design with a secondary rotating auger. *See Figure 3-12.* A two-stage snow thrower has a blower mechanism that adds additional discharge power to the unit for increased capacity and discharge distance. It also has drift cutters for working through deep snowfalls and snowdrifts. Two-stage snow throwers can effectively move a higher volume of snow, maintain higher working speeds, and discharge snow at a greater distance than single-stage snow throwers.

Commercial and Industrial OPE

Commercial and industrial outdoor power equipment is equipment that is used exclusively outdoors in a commercial or industrial environment. Commercial environments include outdoor athletic facilities, office buildings, schools, hospitals, and similar institutions. Industrial environments include construction job sites and any areas undergoing extensive renovations. The most common types of outdoor power equipment used in these locations include portable generators, concrete mixers, and compaction equipment.

SINGLE-STAGE
- Engine (covered)
- Discharge chute
- Single curved blade
- Snow collection hopper

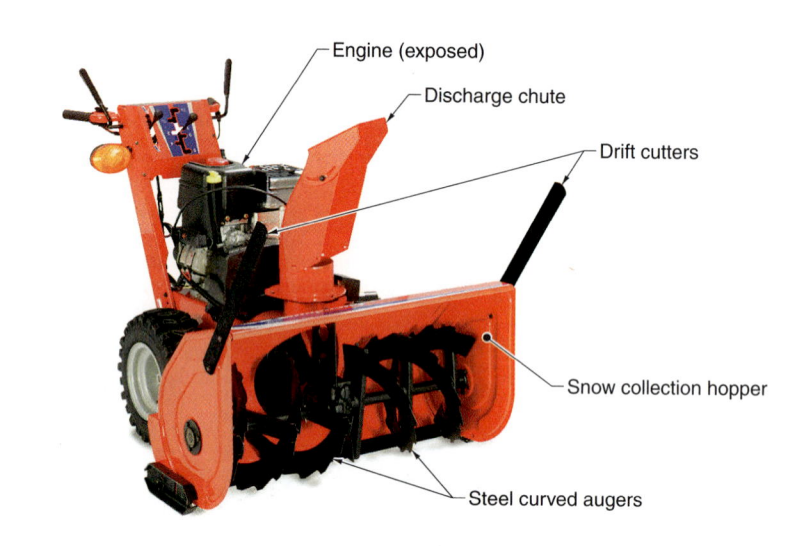

TWO-STAGE
- Engine (exposed)
- Discharge chute
- Drift cutters
- Snow collection hopper
- Steel curved augers

Figure 3-12. Snow throwers can be single-stage or two-stage. Two-stage snow throwers are designed to remove more snow than single-stage types.

TECH FACT
The term "snow blower" is sometimes used to describe snow removal equipment other than snow plows. The correct term is "snow thrower," since most modern snow removal equipment throws snow aside rather than blowing it aside.

Snow Throwers

Portable Generators. A *portable generator* is a portable machine that changes rotating mechanical energy into electric energy. Portable generators are engine-powered applications that create usable electrical power by converting energy provided by the engine into electrical power using an electrical generating component. The operating engine is rigidly connected to an armature. An *armature* is the rotating part of a generator that consists of a segmented iron core surrounded by copper wires wound tightly together. As the armature rotates at engine speed, the copper wire windings in the armature, in conjunction with a stationary magnetic field, induce electricity, which is then made available to the electrical outlet on the generator panel of the generator.

Portable generators are typically used in commercial and industrial environments where permanent electric power is not available to provide electric power for power tools and other electric equipment. Portable generators are often used in residential applications to provide auxiliary electric power for critical loads, such as refrigerators and sump pumps, during utility power outages. *See Figure 3-13.*

PORTABLE GENERATORS

Power cord to components requiring power — Power panel — Engine

Figure 3-13. Portable generators are often used in residential applications to provide auxiliary electric power for critical loads, such as refrigerators and sump pumps, during utility power outages.

34 SMALL ENGINE AND EQUIPMENT MAINTENANCE

Concrete Mixers. A *concrete mixer* is portable or stationary equipment that consists of a rotating drum or paddle used to mix concrete ingredients. Concrete ingredients are introduced into the open end of the concrete mixer and discharged by tilting the drum. Concrete mixers are powered by small four-stroke cycle engines. Most small, portable concrete mixers have trailer hitch mechanisms that allow the concrete mixers to be towed behind vehicles. *See Figure 3-14.*

Concrete mixers mix mortar as well as concrete. With a mixer that mixes concrete, the drive shaft end is a pinion gear that meshes with a large ring gear. The ring gear is typically affixed to the outside diameter of the mixer drum. Inside the drum are stationary paddles that rotate at drum speed, mixing the ingredients. With a mixer that mixes mortar, there is a rotating shaft with paddles within a stationary drum.

For both types of mixers, a flexible drive belt is most commonly used as the connection between the engine and the rest of the machine.

Drive power is applied to the system using simple lever and pulley components, which apply the appropriate amount of belt tension to the system to perform the required work.

Compaction Equipment. Compaction equipment includes vibratory plates, soil compactors, and concrete vibrators. Compaction equipment is used for construction projects where earth has been excavated for construction purposes.

CONCRETE MIXERS

Figure 3-14. Concrete mixers are portable or stationary devices with rotating drums or paddles used to mix concrete ingredients and are typically powered by four-stroke cycle engines.

A *vibratory plate* is gasoline- or diesel-powered compaction equipment used to compact the surface of pavement and/or pavement underlayment materials. A *soil compactor* is a mechanical or manual device used to improve the bearing capacity of soil by tamping or vibrating the soil. A *concrete vibrator* is a power tool used to agitate and consolidate freshly placed concrete and produce close contact with a form or mold.

In general terms, compaction equipment is designed to compress loose or viscous materials into more dense consistencies, allowing the compacted material to support the weight of construction. Compaction equipment is typically small enough to be powered by small engines. For example, most soil compactors have walk-behind designs, similar to lawn mowers and snow throwers. **See Figure 3-15.**

COMPACTION EQUIPMENT

Wacker Nueson Corporation

Figure 3-15. Most soil compactors are powered by four-stroke cycle engines and are used to improve the bearing capacity of soil by tamping or vibrating the soil.

Chapter 3 Quick Quiz®

Additional types of outdoor power equipment that can be powered by small engines include masonry saws, post hole diggers, and log splitters.

Small Engine Fundamentals

4

INTRODUCTION

Although small engines can have two- or four-stroke cycle designs, most small engines that are used to power outdoor power equipment are four-stroke cycle engines. Four-stroke cycle engines operate similarly to other types of internal combustion engines but have unique components and functions. Small engines can have up to seven different systems. Five of these systems are found on every four-stroke cycle small engine.

FOUR-STROKE CYCLE THEORY

The four-stroke cycle engine is the most common type of small engine. A *four-stroke cycle engine* is an internal combustion engine that uses four piston strokes known as intake, compression, power, and exhaust to complete one operating cycle. Gasoline-powered four-stroke cycle engines power many types of equipment, such as lawn mowers, snow throwers, and lawn and garden tractors, by generating the precise amount of power required to perform work. *See Figure 4-1.* Most small engine components can be easily accessed.

FOUR-STROKE CYCLE ENGINE APPLICATIONS

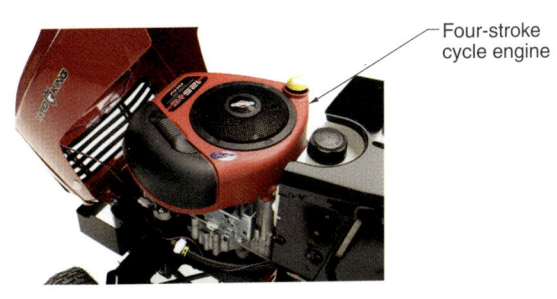

ROTARY LAWN MOWER — Four-stroke cycle engine

SINGLE-STAGE SNOW THROWER — Four-stroke cycle engine

LAWN AND GARDEN TRACTOR — Four-stroke cycle engine

Figure 4-1. Gasoline-powered four-stroke cycle engines typically power lawn mowers, snow throwers, lawn and garden tractors, and many other types of outdoor power equipment.

38 SMALL ENGINE AND EQUIPMENT MAINTENANCE

The main difference between small engines and other types of engines is their small capacity and simplicity of design. Small engines generate modest amounts of power, typically up to 32 HP. In comparison, the engine for a typical automobile generates up to 360 HP. The size of a small engine and its design allow for easier operation, maintenance, and repair than larger engines.

Small engines are commonly designed for simple tasks such as cutting lawns and turning soil. *See Figure 4-2.* Unlike engines for automobiles and other vehicles that frequently accelerate, decelerate, or idle for long time periods, small engines typically run at constant speed. They change speed in response to modest changes in load such as when a lawn mower is pushed over a patch of thick grass.

Also, unlike automotive engines, small engines do not have to be sized to fit under an automobile hood or allow for connection to computers and other devices. Their compact size also makes parts simpler to install, adjust, and remove.

Four-Stroke Cycle Engine Operation

A four-stroke cycle engine uses four piston strokes to complete one operating cycle. The four-stroke cycle has four distinct piston actions that occur in succession. The actions include *intake*, *compression*, *power*, and *exhaust* actions. The piston makes two complete passes in the cylinder. The entire process involves two up strokes and two down strokes of the piston.

SMALL ENGINE DESIGNS

FOR HANDHELD EQUIPMENT

FOR LAWN MOWERS

FOR SNOW THROWERS

FOR LAWN AND GARDEN TRACTORS

FOR COMMERCIAL EQUIPMENT

Figure 4-2. Small engines are designed to perform simple tasks with specific types of outdoor power equipment.

Five events are completed in one operating cycle. An operating cycle includes intake, compression, ignition, power, and exhaust events. It requires two revolutions (720°) of the crankshaft. *See Figure 4-3.* A four-stroke cycle engine operates through the following procedure:

1. The rewind starter cord is pulled, or electric starter pushbutton depressed, causing the engine to rotate and allowing precise amounts of fuel and filtered air to mix in the carburetor.

Chapter 4—Small Engine Fundamentals 39

2. The fuel and filtered air mixture rushes into the engine to be compressed, ignited, and burned in a controlled process known as internal combustion. Internal combustion produces hot rapidly expanding gases.

3. As the hot gases produced by internal combustion expand, they push the piston down, which is a smooth, well-lubricated cylindrical component.

4. The piston then drives the crank shaft, which rotates a blade or performs other work.

5. Valves allow air and fuel into the combustion chamber above the piston and allow spent gases to exit through the exhaust system.

FOUR-STROKE CYCLE ENGINE OPERATION

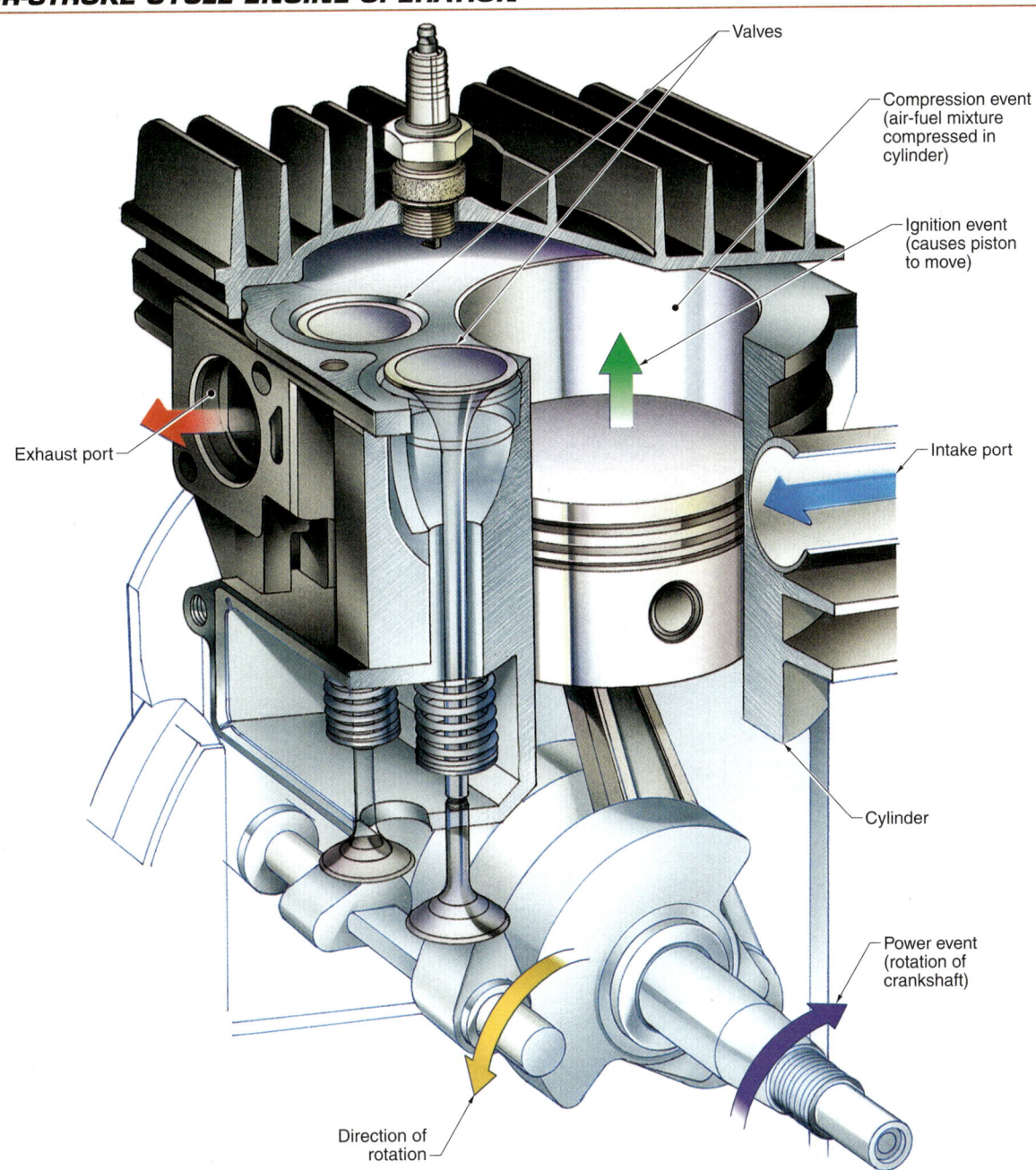

Figure 4-3. A four-stroke cycle engine completes five events in one operating cycle, including intake, compression, ignition, power, and exhaust events.

The process is designed to become self-sustaining from the time the engine starts until the moment it stops. Timed electrical surges cause the spark plug to fire repeatedly inside the combustion chamber. This ignites each fresh supply of air and fuel and produces gases that continually drive the piston and crankshaft. In most engines, including engines found in automobiles and outdoor power equipment, combustion occurs in a four-stroke cycle.

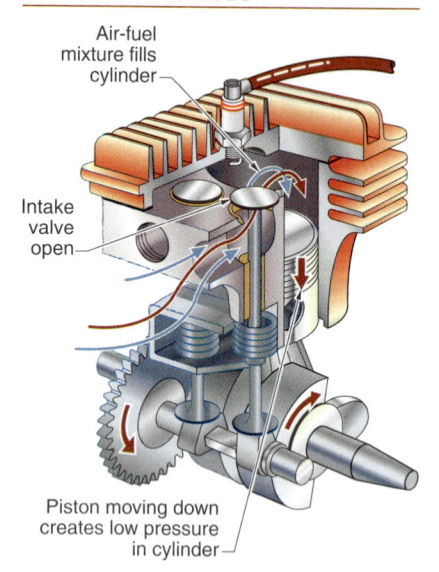

Intake Event. An *intake event* is an engine operation event in which an air-fuel mixture is introduced to a combustion chamber. *See Figure 4-4.* The mixture of air and fuel is introduced into the combustion chamber during an intake stroke. The intake valve is open and the piston moves from top dead center (TDC) to bottom dead center (BDC).

The effect of the piston is similar to the suction produced by a syringe drawing liquid. For example, as the plunger inside the syringe slides toward the handle, it creates a low-pressure area at the tip. Similarly, as the piston moves toward BDC, it creates a low-pressure area in the cylinder and draws the air-fuel mixture through the open intake valve. The mixture continues to flow, due to inertia, as the piston moves beyond BDC. Once the piston moves a few degrees beyond BDC, the intake valve closes, sealing the air-fuel mixture inside the cylinder.

Compression Event. A *compression event* is an engine operation event in which a trapped air-fuel mixture is compressed inside a combustion chamber. *See Figure 4-5.* Compression occurs as the piston travels toward TDC. The air-fuel mixture is compressed for a more efficient burn and to allow energy to be released faster when the mixture is ignited. The engine is able to do work because the energy required for compression—and stored in the flywheel—is far less than the force produced during combustion. In a typical small engine, compression requires one-fourth the energy produced during combustion. The surplus energy drives the power stroke.

INTAKE EVENT

Figure 4-4. An intake event is an engine operation event in which an air-fuel mixture is introduced to a combustion chamber.

COMPRESSION EVENT

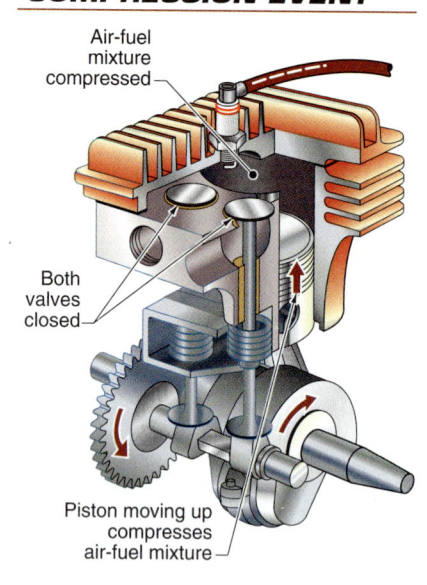

Figure 4-5. A compression event is an engine operation event in which a trapped mixture of air and fuel is compressed inside the combustion chamber.

Ignition Event. An *ignition (combustion) event* is an engine operation event in which fuel is ignited to release heat energy. *See Figure 4-6. Combustion* is a rapid, oxidizing chemical reaction in which a fuel chemically combines with atmospheric oxygen and releases energy in the form of heat. Proper combustion involves a short but finite time to spread a flame throughout a combustion chamber. The spark at the spark plug initiates combustion at approximately 20° of crankshaft rotation before top dead center (BTDC).

The oxygen and fuel vapor are consumed by a progressing flame front. A *flame front* is a boundary wall that separates the charge from combustion by-products. The flame front progresses across the combustion chamber until the entire charge has burned.

TECH FACT
The first 50,000,000 Briggs & Stratton engines were produced between 1924 and 1967 (43 years). The second 50,000,000 Briggs & Stratton engines were produced between 1968 and 1975 (7 years).

IGNITION EVENT

Spark initiates combustion

Both valves closed

① SPARK OCCURS JUST BEFORE TDC

② CHARGE BEGINS BURNING

③ FLAME FRONT SPREADS THROUGHOUT COMBUSTION CHAMBER

④ FLAME FRONT COMPLETES BURN

Figure 4-6. An ignition event is an engine operation event in which a charge is ignited and rapidly oxidized through a chemical reaction to release heat energy.

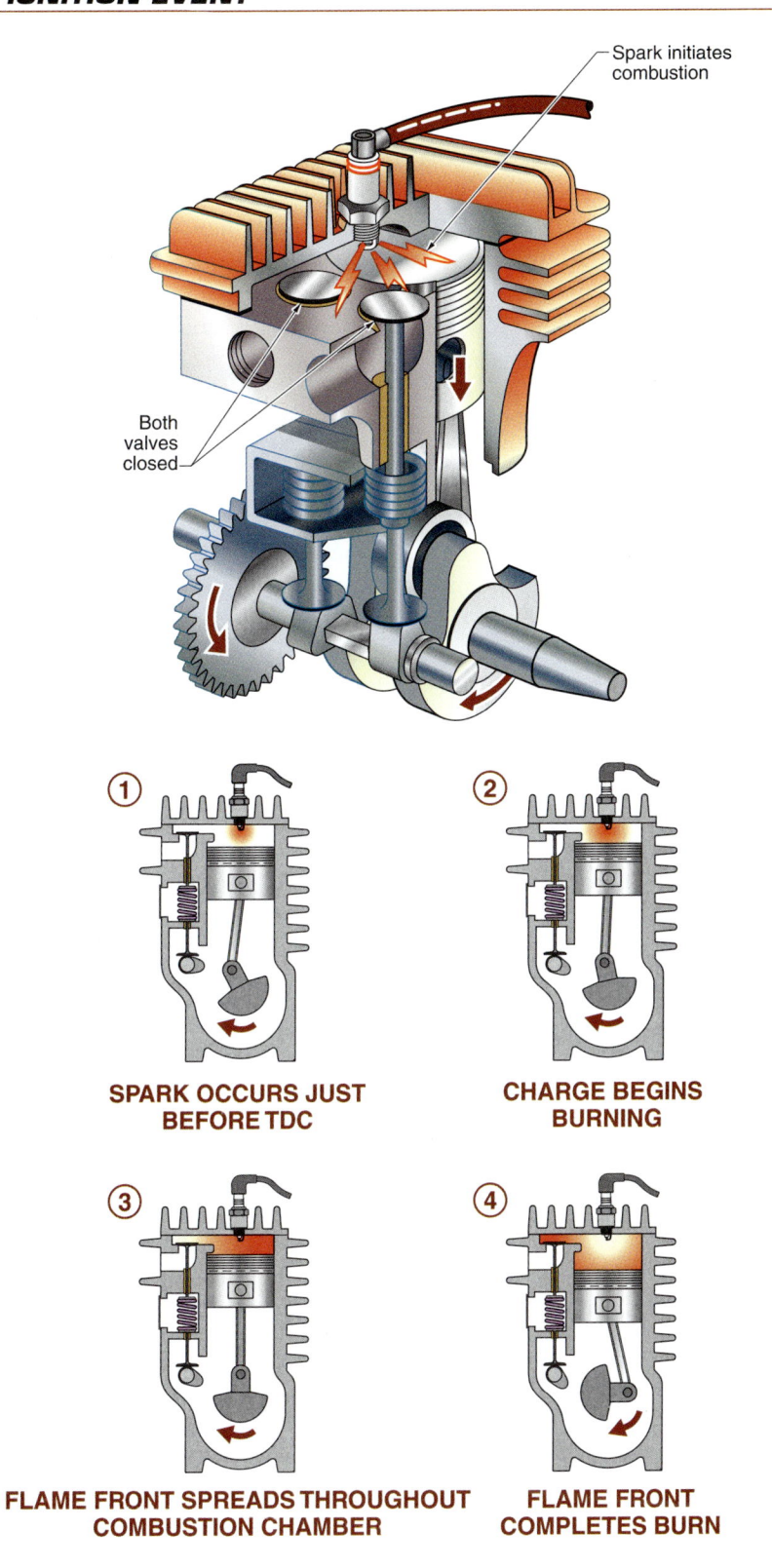

Power Event. A *power event* is an engine operation event in which a compressed charge is ignited and hot rapidly expanding gases force a piston head away from a cylinder head. *See Figure 4-7.* The intake and exhaust valves of the engine are closed. At approximately 20° BTDC, the spark plug initiates combustion, creating a flame that burns the compressed air-fuel mixture. The hot gases produced by combustion have no way to escape, so they push the piston away from the cylinder head. The motion is transferred through the connecting rod to apply torque to the crankshaft.

POWER EVENT

Charge ignited

Both valves closed

Connecting rod

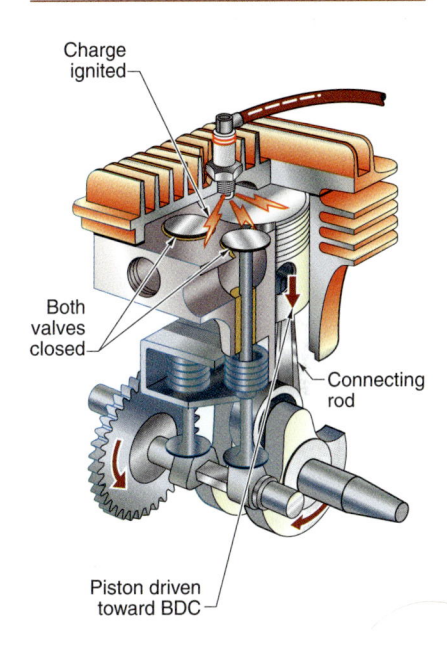

Piston driven toward BDC

Figure 4-7. A power event is an engine operation event in which a compressed charge is ignited and hot expanding gases force a piston head away from a cylinder head.

Exhaust Event. An *exhaust event* is an engine operation event in which spent gases are removed from a combustion chamber and released into the atmosphere. *See Figure 4-8.* As the piston reaches BDC during the power stroke, the power stroke is completed. The exhaust valve opens, allowing the piston to evacuate exhaust as it moves, once again, toward TDC. With the chamber cleared of exhaust, the piston reaches TDC. An entire cycle is complete.

EXHAUST EVENT

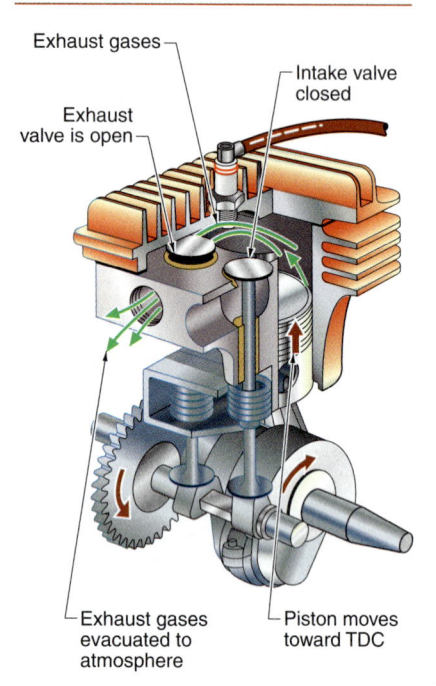

Figure 4-8. An exhaust event is an engine operation event in which spent gases are removed from the combustion chamber and released to the atmosphere.

Four-Stroke Cycle Events

The Two-Stroke Cycle Alternative

A two-stroke cycle engine is an internal combustion engine that uses two distinct piston strokes to complete one operating cycle of the engine. The crankshaft turns only one revolution for each complete operating cycle. The two-stroke cycle engine provides twice as many power strokes in the same number of crankshaft rotations as a four-stroke cycle engine. Like a four-stroke cycle engine, a two-stroke cycle engine completes five events in one operating cycle. However, some events occur concurrently, such as ignition/power and exhaust/intake.

The ignition/power event occurs when the piston moves toward TDC and the compressed charge in the cylinder is ignited. During this time, the crankcase has already filled with a fresh air-fuel mixture. When the charge is ignited, expansion of hot combustion gases force the piston toward BDC. Piston motion is transferred from the piston through the connecting rod, causing the crankshaft to rotate. When the piston moves toward BDC, the exhaust port is uncovered and exhaust gases are discharged through the side of the cylinder.

The exhaust/intake event occurs as the piston continues moving toward BDC. The intake port opens and the air-fuel mixture is routed into the cylinder. The shape of the piston head helps to divert the incoming air-fuel mixture to the top of the cylinder. This prevents incoming air and fuel from passing across the top of the piston and out the exhaust port without burning. The piston acts as a valve, exposing the intake and exhaust ports at designated moments in the cycle.

Two-stroke cycle engines are widely used for chain saws, leaf blowers, and other handheld equipment. In the past, two-stroke cycle engines were preferred for handheld equipment because of their lightweight design. However, the latest technology has reduced the weight of four-stroke cycle engine components, creating the potential for inroads in the handheld equipment industry. Two-stroke cycle engines are also used for outboard motors and motorcycles. *Note:* They are no longer used on street bikes in many countries because of the higher emissions they produce.

Chapter 4—Small Engine Fundamentals   43

TWO-STROKE CYCLE ENGINE EVENTS

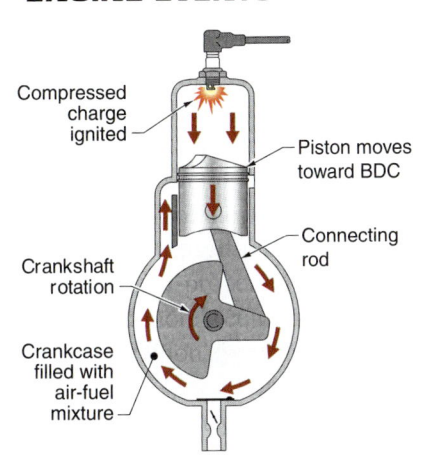

IGNITION/POWER
- Compressed charge ignited
- Piston moves toward BDC
- Connecting rod
- Crankshaft rotation
- Crankcase filled with air-fuel mixture

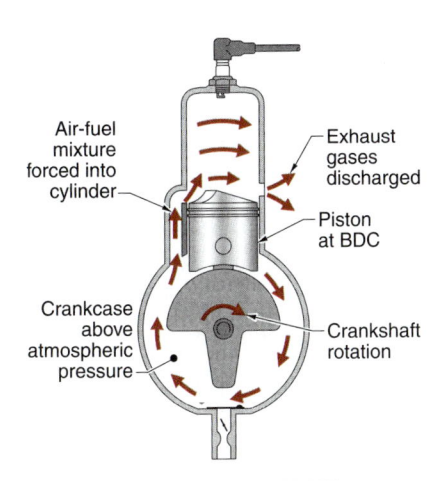

EXHAUST/INTAKE
- Air-fuel mixture forced into cylinder
- Exhaust gases discharged
- Piston at BDC
- Crankcase above atmospheric pressure
- Crankshaft rotation

COMPRESSION
- Compressed charge
- Piston moves toward TDC
- Both ports closed
- Crankcase below atmospheric pressure
- Air-fuel mixture drawn into crankcase
- Crankshaft rotation

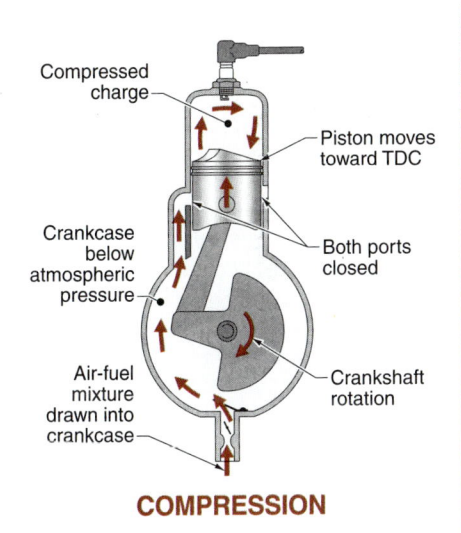

Overhead Valves: Gaining Popularity in Small Engines

Placing valves next to the piston is just one method of engine configuration. Automotive engineers determined long ago that a significant advantage could be gained in many high-horsepower engines by installing valves in the cylinder head so that they face the piston. Pivoting rocker arms moved by push rods open the valves. One of the main advantages of overhead valve (OHV) design is a more symmetrical combustion chamber, resulting in a more efficient burning of the air-fuel mixture. Another advantage of overhead valve design is improved cooling of the combustion chamber. The overhead valve design provides increased surface area, which allows a greater volume of air to pass over the cooling fins, for transferring heat energy to the atmosphere. Briggs & Stratton designs and manufactures many different small engines that use overhead valves. The letters "OHV" imprinted on the shroud of the engine indicates the use of overhead valves.

Overhead Valve Engines

Engine Components and Functions

All small engines have the same basic components that perform the same functions. These components are used for starting, running, and stopping the engine. *See Figure 4-9.* Small engines include the following basic components:

- **fuel tank:** A *fuel tank* is a liquid storage vessel commonly connected to an engine for the purpose of holding fuel, such as gasoline or diesel fuel, to power the engine. Fuel tanks can be composed of sheet metal or a polymer and, depending on the size of the engine, can hold ½ gal. to 5 gal. (1.8 L to 19 L) of fuel.

- **shroud:** A *shroud* is a rigid, protective cover attached to an engine near the fuel tank to protect the engine from moisture, dust, dirt, and other debris.

- **rewind cord:** A *rewind cord* is a device that is pulled to start the combustion process within an engine ("start the engine"). On some models, a starter motor replaces the rewind cord and uses a power cord connected to an electrical outlet (AC power) or a battery (DC power) to start the engine.

- **blower housing:** A *blower housing* is a sheet metal or composite, such as plastic, component that encompasses a fan to direct cooling air to a cylinder block and cylinder head.

- **flywheel:** A *flywheel* is a cast iron, aluminum, or zinc disk that is mounted on one end of a crankshaft to provide inertia for an engine to prevent the loss of engine speed between combustion intervals. The flywheel is located under the blower housing.

- **flywheel magnet:** Rotating magnets that are integrated in the flywheel work in conjunction with the ignition system (ignition armature and spark plug) to produce a spark in the combustion chamber.

- **primer bulb:** A *primer bulb* is an enrichment system consisting of a rubber bulb filled with fuel or air connected to a fuel bowl by a passageway. Wet primer bulbs are filled with fuel, while dry primer bulbs are filled with air.

- **air cleaner:** An *air cleaner* is a device designed to remove airborne impurities, such as dust and dirt, from the air. Air cleaners on small engines are typically composed of paper or plastic foam.

- **carburetor:** A *carburetor* is a device that draws in fuel from a fuel tank and outside air to form a combustible vapor that is fed into the combustion chamber of an engine.

- **valve:** Intake and exhaust valves open and close at precisely timed intervals to allow air and fuel to enter the engine combustion chamber and to allow spent gases to exit.

- **engine block:** An *engine block* is the main structure of an engine that supports and helps maintain the alignment of internal and external components.

- **piston:** A *piston* is a cylindrical engine component that slides back and forth in a cylinder bore by forces produced during a combustion process. The piston is pushed through the cylinder by the force of expanding gases. The motion of the piston causes the crankshaft to turn. Momentum then carries the piston back toward the top of the cylinder.

- **engine oil:** Engine oil circulates through an operating engine to lubricate key components such as the piston and crankshaft. It also provides generalized cooling by drawing heat away from internal engine surfaces. Engine oil is added to the engine directly through the oil fill area. Engine oil level is checked with a metal or plastic dipstick. It is stored in the crankcase when the engine is not in operation.

- **flywheel brake:** A *flywheel brake* is a device that is included on the engine of equipment that requires constant operator presence such as rotary lawn mowers. It is equipped with a stop switch. The two components are designed to stop the engine if the operator presence controls are released.

- **muffler:** A *muffler* is an engine component fitted with baffles and plates that subdues noise produced from exhaust gases exiting a combustion chamber.

- **air vane:** An *air vane* is a device that monitors engine RPM so that a governor can maintain the selected engine speed.

TECH FACT

Some engines use a stop contact to ground a spark plug and stop the engine. By pressing a lever, the operator creates contact between the spark plug terminal and the engine chassis to stop the engine. A stop contact ensures that the engine will not start accidentally. For complete protection, the spark plug lead should always be disconnected and secured away from the plug during maintenance.

Chapter 4—Small Engine Fundamentals

- **cooling fin:** A *cooling fin* is an integral thin cast strip designed to provide efficient air circulation and the dissipation of heat away from an engine cylinder block into the air stream. Cooling fins help reduce engine temperatures when air circulates across hot engine surfaces.
- **cylinder head:** A *cylinder head* is a cast aluminum alloy or cast iron engine component fastened to the end of the cylinder block farthest from a crankshaft.
- **crankshaft:** A *crankshaft* is an engine component, such as a lawn mower blade or snow thrower auger, that converts the linear (reciprocating) motion of a piston and connecting rod into rotary motion.
- **crankcase:** A *crankcase* is an engine component that houses and supports a crankshaft.
- **spark plug:** A *spark plug* is a component that isolates electricity induced in secondary windings and directs a high-voltage charge to the spark gap at the tip of a spark plug.

A *spark gap* is the distance from the center electrode to the ground electrode of a spark plug. A spark plug shell secures the spark plug in the engine. The spark plug lead (wire) is connected to the spark plug terminal. An insulator prevents shorting between the center electrode and the grounding electrode. The high-voltage spark jumps across the spark gap, igniting the air-fuel mixture in the combustion chamber.

SMALL ENGINE COMPONENTS . . .

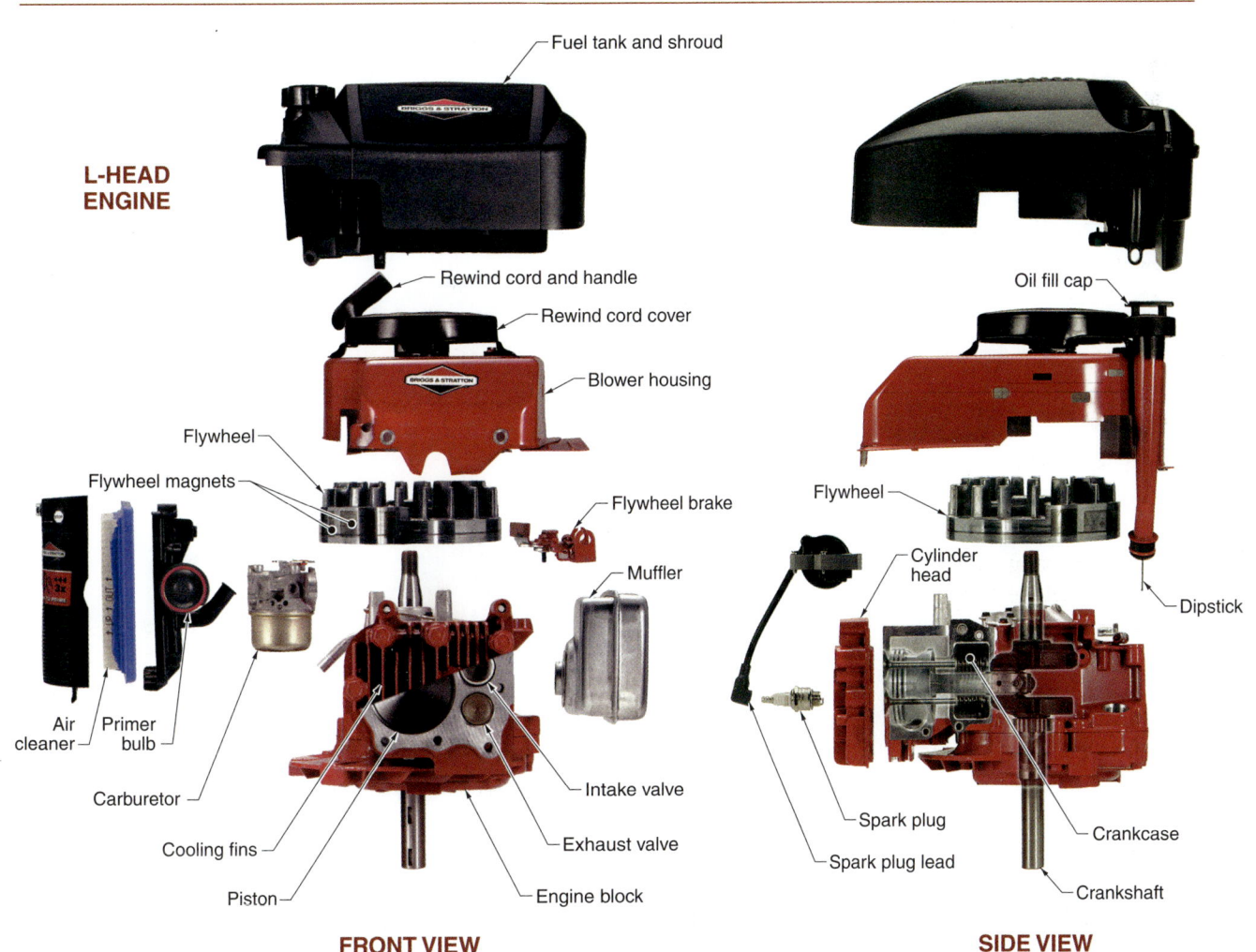

Figure 4-9. Basic small engine components are used for starting, running, and stopping the engine.

46 SMALL ENGINE AND EQUIPMENT MAINTENANCE

. . . SMALL ENGINE COMPONENTS . . .

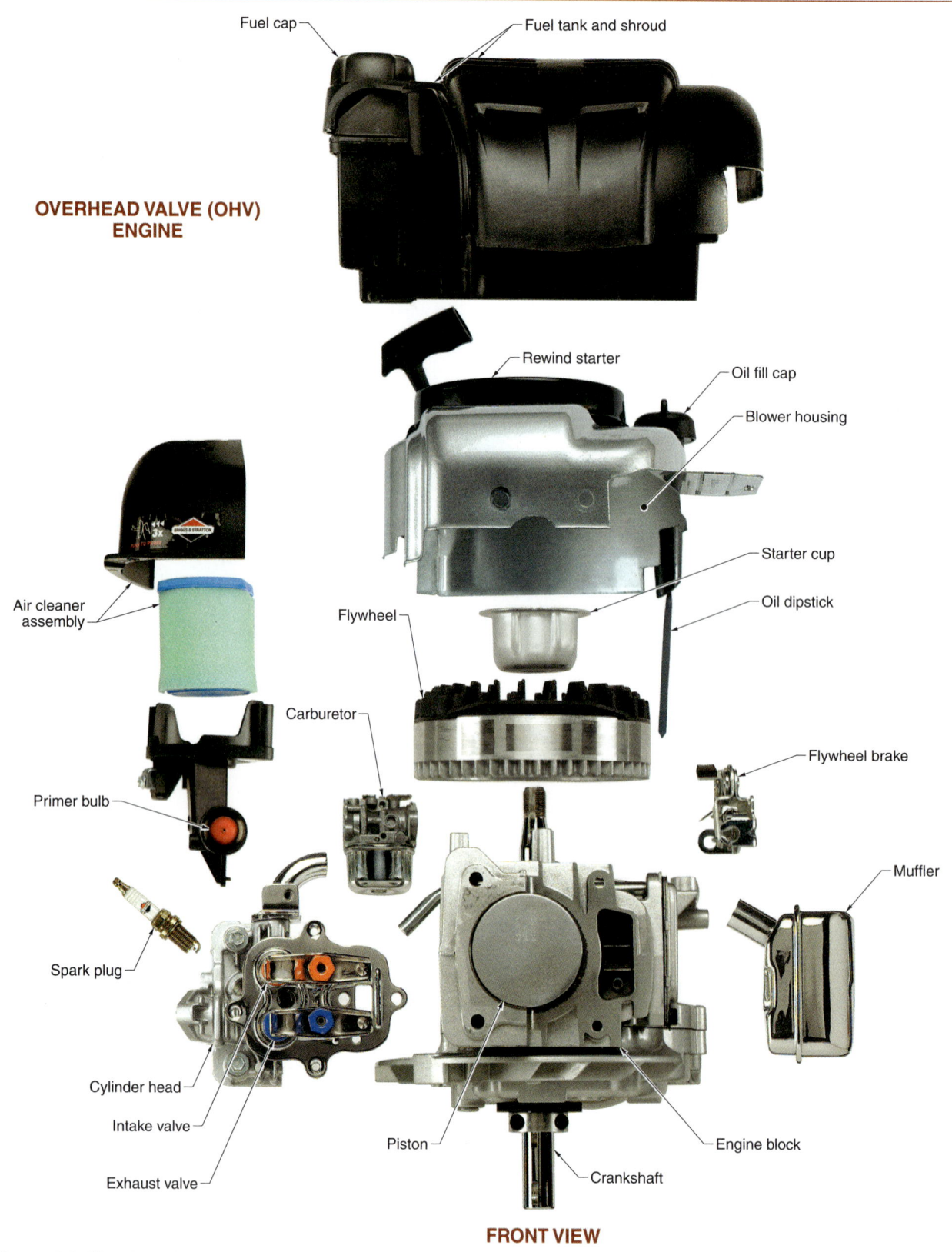

OVERHEAD VALVE (OHV) ENGINE

- Fuel cap
- Fuel tank and shroud
- Rewind starter
- Oil fill cap
- Blower housing
- Starter cup
- Oil dipstick
- Air cleaner assembly
- Flywheel
- Carburetor
- Flywheel brake
- Primer bulb
- Muffler
- Spark plug
- Cylinder head
- Intake valve
- Exhaust valve
- Piston
- Crankshaft
- Engine block

FRONT VIEW

Figure 4-9. (Continued)

Chapter 4—Small Engine Fundamentals **47**

... SMALL ENGINE COMPONENTS

Fuel tank and shroud

Rewind starter

Oil fill cap

Blower housing

Starter cup

Flywheel

Ignition armature

Oil dipstick

Spark plug lead

Piston

Crankcase

Valve springs

Push rods

Cylinder head

Connecting rod

Crankshaft

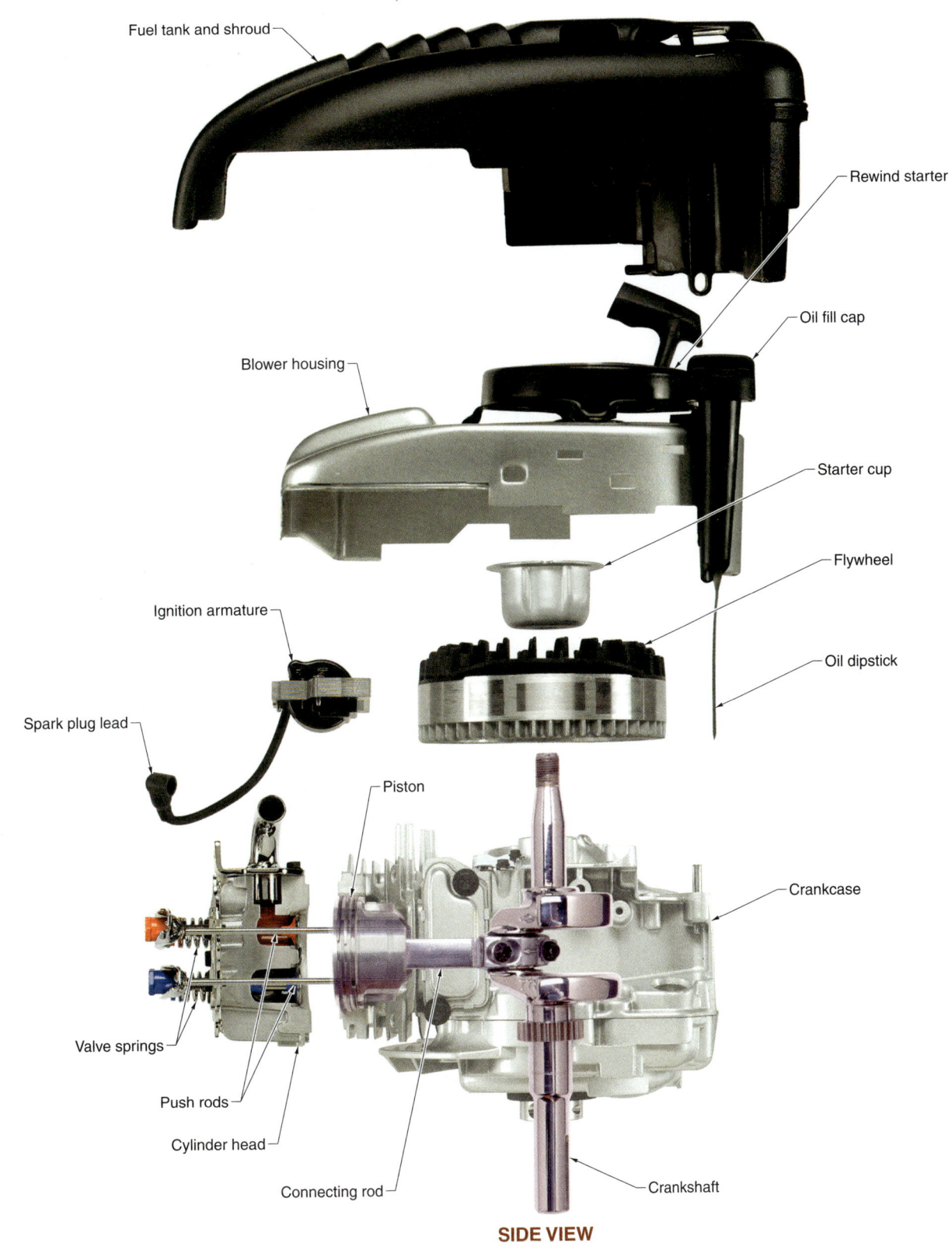

SIDE VIEW

Figure 4-9. (Continued)

SMALL ENGINE SYSTEMS

There are up to seven working systems in small engines: compression, fuel, ignition, lubrication and cooling, governor (speed control), electrical, and braking. Five systems, which are used on four-stroke cycle engines, generate the power to rotate a blade, turn a wheel, or perform other work. The other two systems, electrical and braking, are commonly included to improve safety and convenience. For example, braking systems are typically found on rotary lawn mowers.

Compression Systems

A compression system includes a number of valves, a piston and cylinder bore, a number of piston rings, and a crankcase. Twin-cylinder engines for heavy-duty equipment contain two cylinders, each with a separate piston and valves.

The designers of the first internal combustion engines discovered that fuel burns more efficiently when compressed in a sealed chamber prior to being burned. Compression of the air-fuel mixture in a small engine begins as the intake valve closes. Trapped vapors are pushed toward the cylinder head by the piston and compressed into a space about one-sixth to one-eighth its original volume. The amount of compression is an indicator of the efficiency of an engine and is why a tightly sealed combustion chamber is a requirement for the best engine performance.

Valves. A *valve* is an engine component that opens and closes at precise times to allow the flow of an air-fuel mixture into and exhaust gases from a cylinder. Valves located in the combustion chamber allow fuel vapors and air to enter the cylinder and allow exhaust gases to exit at precisely timed intervals.

A typical four-stroke cycle small engine contains one intake valve and one exhaust valve per cylinder. Most small engines have one cylinder and use an L-head (or flathead) design. The valves are installed in a valve chamber next to the piston. However, overhead valve (OHV) designs offer greater efficiency and are increasingly popular with consumers. The valves are located in the cylinder head directly in line with the piston and are actuated by pivoting rocker arms.

The location of valves determines the type of head design and the necessary components for the valve train. In an L-head engine, the valves are located in the cylinder block on one side of the cylinder. Although overhead valve and direct overhead valve (DOV) engines are installed on some modern types of OPE, most small engines are L-head engines. *See Figure 4-10.*

FOUR-STROKE CYCLE ENGINE VALVING SYSTEMS

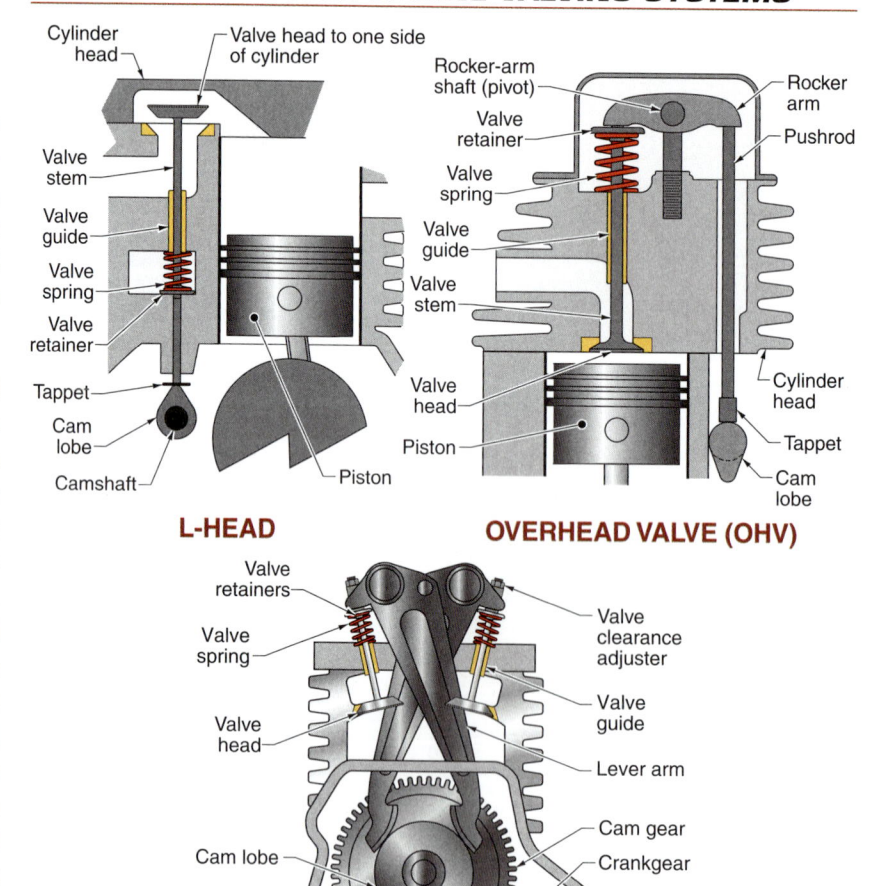

Figure 4-10. Valve location determines whether an engine is an L-head or overhead valve engine.

PISTONS

Pistons. A *piston* is a cylindrical engine component that slides back and forth in a cylinder bore by forces produced during the combustion process. *See Figure 4-11.* The piston rides through a cylinder bore, much as a plunger rides through the chamber in a hand-operated air pump. At the appropriate moment, the cylinder bore is sealed so that the air-fuel mixture is compressed as the piston moves toward the cylinder head. When the mixture is ignited, rapidly expanding gases force the piston back down through the cylinder bore.

Piston Rings. A *piston ring* is an expandable split ring used to provide a seal between a piston and cylinder bore. The space between the outside diameter (OD) of the piston ring and the inside diameter (ID) of the cylinder is narrow enough to permit a thin coating of lubricating oil.

Flexible piston rings, which are installed in grooves in the piston, work together with the lubricating oil to create a seal between the piston OD and the cylinder ID. A good seal between these two surfaces ensures good compression. As the piston is pushed down through the cylinder by expanding gases, a connecting rod transfers the force of those gases to the crankshaft. The momentum from the flywheel perpetuates the four-stroke cycle of the engine.

The crankshaft then converts reciprocating motion to rotational motion. A flywheel is used to continue the momentum through the four-stroke cycle of the engine.

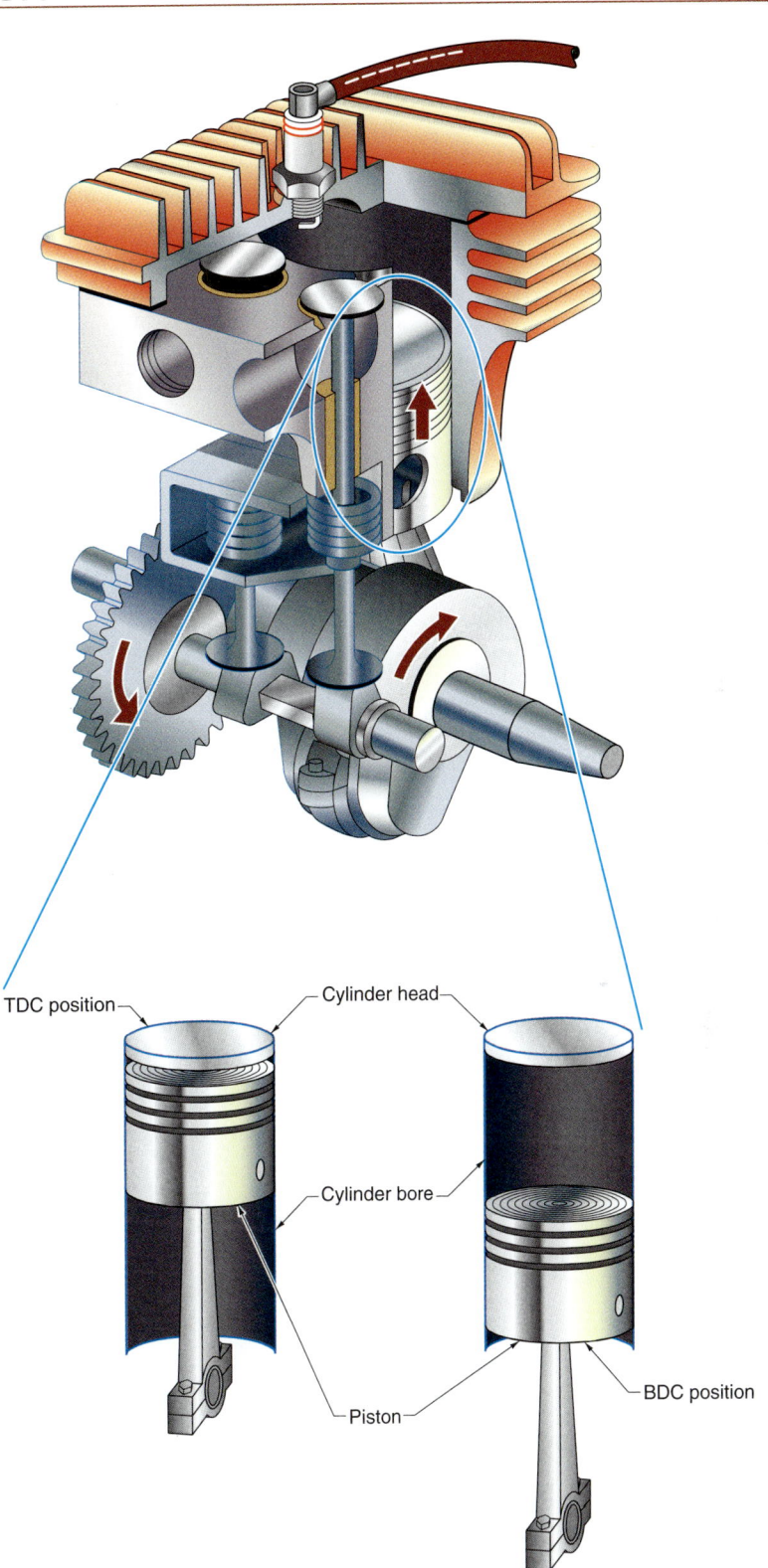

Figure 4-11. A piston acts as the movable end of the combustion chamber by using the forces and heat created during engine operation.

The piston rings commonly used on small engines include a compression ring, wiper ring, and oil control ring. A compression ring is a piston ring located in the ring groove closest to a piston head. The compression ring seals the combustion chamber against any leakage that occurs during the combustion process. A wiper ring is a piston ring with a tapered face located in the ring groove between the compression ring and oil ring.

The wiper ring is used to further seal the combustion chamber and to wipe the cylinder wall clean of excess oil. Combustion gases that pass by the compression ring are stopped by the wiper ring. The oil control ring provides a precise amount of oil for the compression and wiper ring to ride on. The oil control ring also provides an exit for the excess oil on the cylinder wall to return to the crankcase. *See Figure 4-12.*

Crankcases. A *crankcase* is an engine component that houses and supports a crankshaft. In a four-stroke cycle engine, the crankcase also acts as an oil reservoir for the lubrication of engine components. The crankcase may be a part of the engine block or a separate component. Some crankcases consist of multiple parts such as a sump or crankcase cover. *See Figure 4-13.*

PISTON RINGS

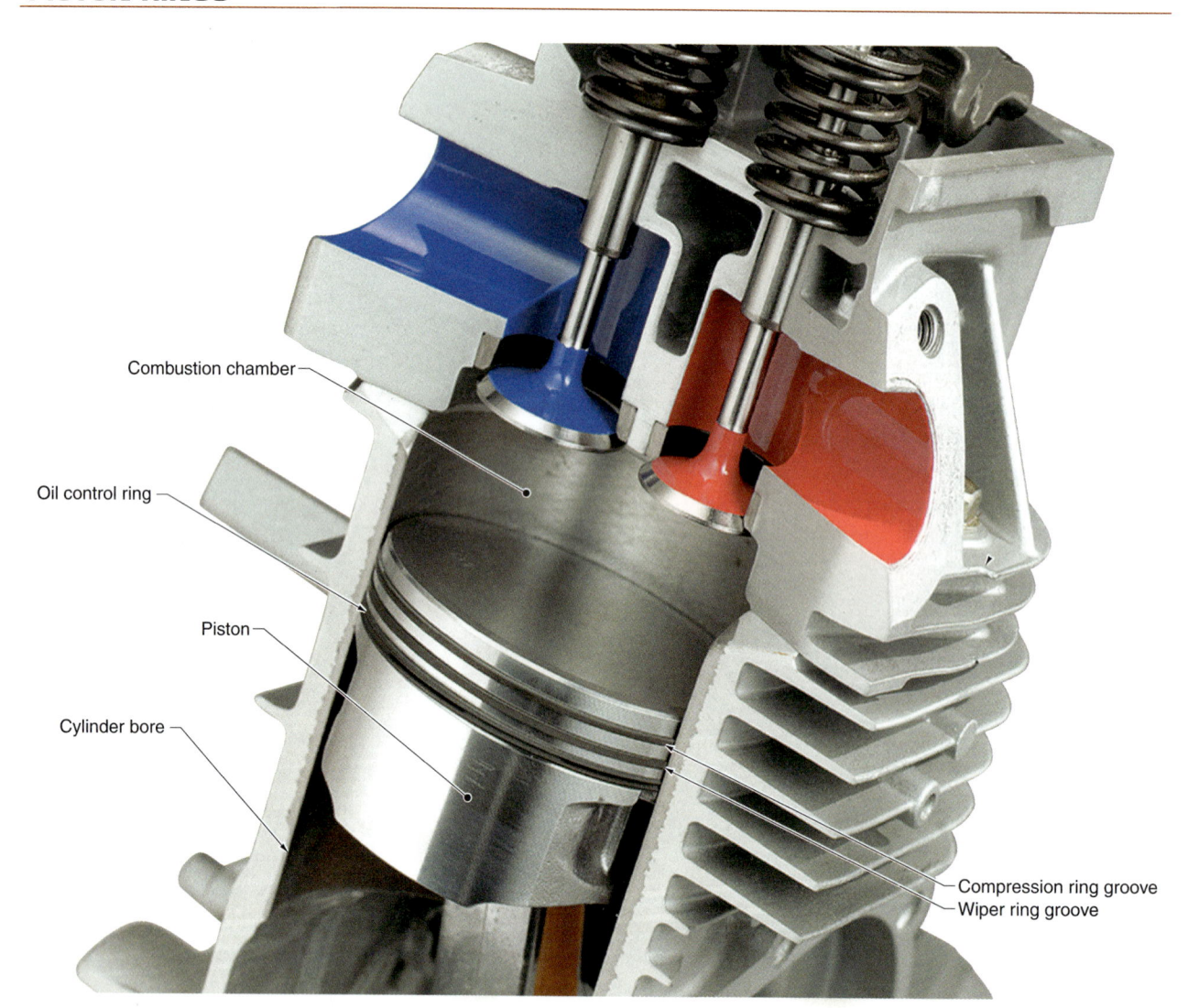

Figure 4-12. Piston rings provide a seal between the piston and the cylinder bore and commonly include a compression ring, wiper ring, and oil control ring.

CRANKCASES

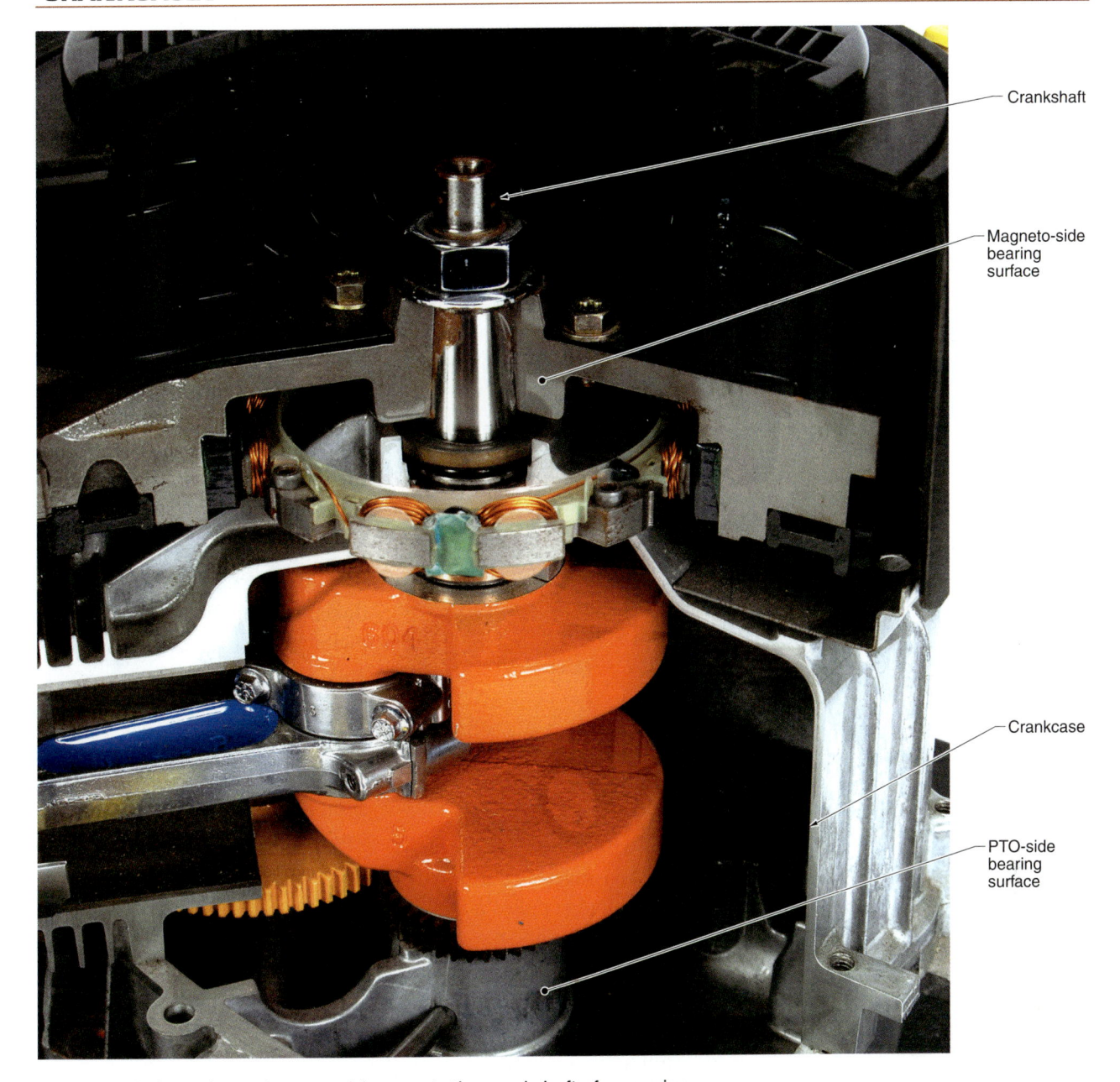

Figure 4-13. A crankcase houses and supports the crankshaft of an engine.

Fuel Systems

A fuel system typically includes a fuel tank, fuel pump, fuel filter, carburetor, and fuel line. To burn, gasoline must be converted from a liquid to a vapor. It must first be converted to a vapor. The vapors that burn in a small engine are formed from a mixture of air and fuel. The correct amount of air and fuel must be mixed together to maintain the required engine speed. The best method used to locate the components of a fuel system is to begin at the fuel tank and trace back through the system. *See Figure 4-14.* The repairs most commonly made for a small engine are on the fuel system.

TECH FACT
When repairing and maintaining any small engine, the replacement engine parts to be installed must be exactly the same as the originals. Installing a similar rather than exact part can prevent the engine from starting or cause it to run poorly.

FUEL SYSTEMS

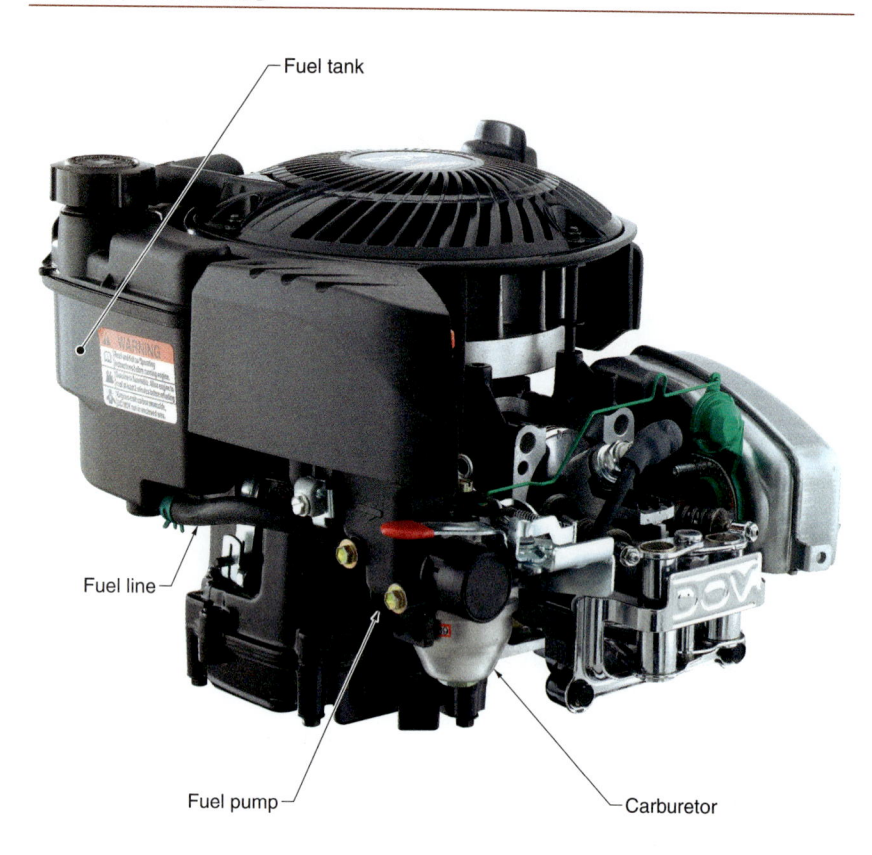

Figure 4-14. A typical small engine fuel system includes a fuel tank, fuel pump (on some models), fuel filter, carburetor, and fuel line.

Note: On outdoor power equipment older than 20 years, the fuel tank is most likely made of steel. Fuel tanks on equipment less than 20 years old are typically made from seamless, blow-molded polymer and are one-piece units. Also, they are commonly covered with a plastic shroud to protect engine parts.

Carburetors. A *carburetor* is an engine component that provides the required air-fuel mixture to a combustion chamber based on engine operating speed and load. A *throat* is the main passageway of a carburetor that directs an air-fuel mixture and air from the atmosphere to a combustion chamber. A *venturi* is the narrow portion of a throat. In essence, a carburetor is a passageway that draws in air and fuel and supplies a mixture of the two to the cylinder.

It is the fully sealed, low-pressure area above the piston that causes the carburetor to draw in the two components the engine requires for combustion. Carburetion occurs when air speed increases at the venturi and air pressure drops. Since fluids flow from areas of high pressure to low pressure, fuel from the bowl or tank is drawn into the throat, mixing with air to form a combustible vapor. **See Figure 4-15.**

To locate the carburetor on a small engine, the fuel tank is first identified. Once the fuel tank is located, the fuel line is identified. The fuel line is a hose connected to one side of the fuel tank. The fuel line carries fuel to the carburetor. On most engines, the force of gravity carries fuel through the fuel line. However, if the fuel tank is mounted low on the engine, gravity alone may not have enough force to move the fuel through the line.

For engines with this design, a fuel pump that uses low pressure in the crankcase to pump fuel is required. The fuel pump is located between the tank and the carburetor or in the carburetor itself. On some engines, the need for either a fuel line or fuel pump is eliminated by having the carburetor mounted directly on the fuel tank and using a pick-up tube in the tank to draw fuel.

On most engines, fuel flows through the fuel line into the fuel bowl of the carburetor. A fuel bowl is a reservoir where a float, similar to the float ball in a toilet tank, regulates the fuel level. From there, a metering device, referred to as a "jet," allows fuel into the emulsion tube inside the pedestal where air and fuel first mix. Fuel travels through the emulsion tube to the venturi, where further mixing with air occurs. If the carburetor is a tank-mounted type, fuel from the tank may be supplied directly to the emulsion tube, without the need for a float. *Note:* Models older than 20 years may include an adjustable jet, while newer models most commonly include nonadjustable fixed jets.

Carburetor Function

CARBURETORS

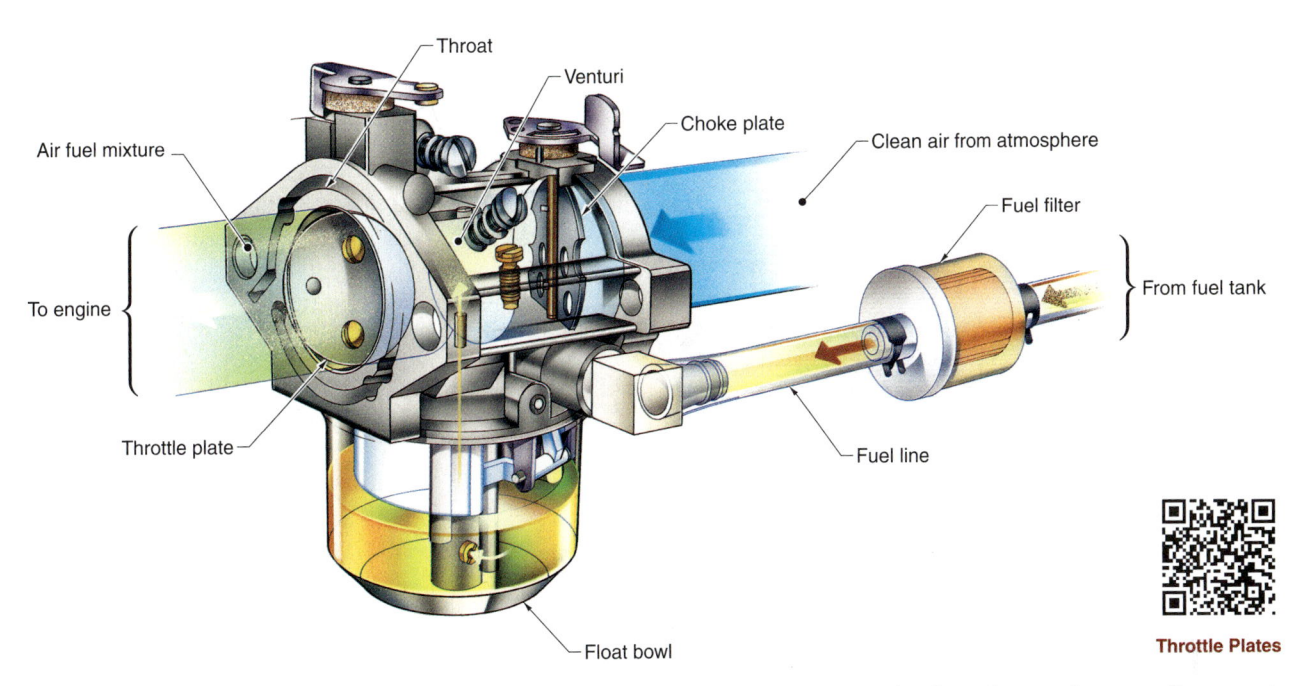

Figure 4-15. A carburetor provides the required air-fuel mixture to a combustion chamber based on engine operating speed and load.

Throttle Plates

At one end of the carburetor throat is a throttle plate. A *throttle plate* is a disk that pivots on a movable shaft to regulate the air and fuel flow in a carburetor. The throttle plate is connected to a control lever (often referred to as a throttle) and opens or closes to increase or decrease engine speed. As the throttle plate opens, air is drawn into the carburetor. Airflow, in turn, determines how much fuel is delivered for combustion.

Many carburetors have an idle speed screw and idle mixture screw. An idle speed screw stops the throttle from closing too much at low speed. An idle mixture screw increases or decreases the amount of fuel allowed into the combustion chamber at idle speeds and at top no-load speeds.

A *choke plate* is a flat plate placed in a carburetor body between the throttle plate and air intake that is used to restrict airflow to help start a cold engine.

A choke plate or primer helps to start an engine in cold temperatures by increasing the amount of liquid fuel, thus increasing the available fuel vapor in the cold combustion chamber, helping the engine to start more easily. The control mechanism for the choke plate is located in the throat between the air filter and the throttle plate. Closing the choke reduces airflow. Low pressure created inside the engine keeps the fuel flowing. The use of the choke enriches the mixture. Although it is not effective for running an engine, this method helps to start a cold engine. Once the engine starts for a moment, the choke should be opened to allow the engine to operate normally. **See Figure 4-16.**

THROTTLES AND CHOKE PLATES

Choke Plates

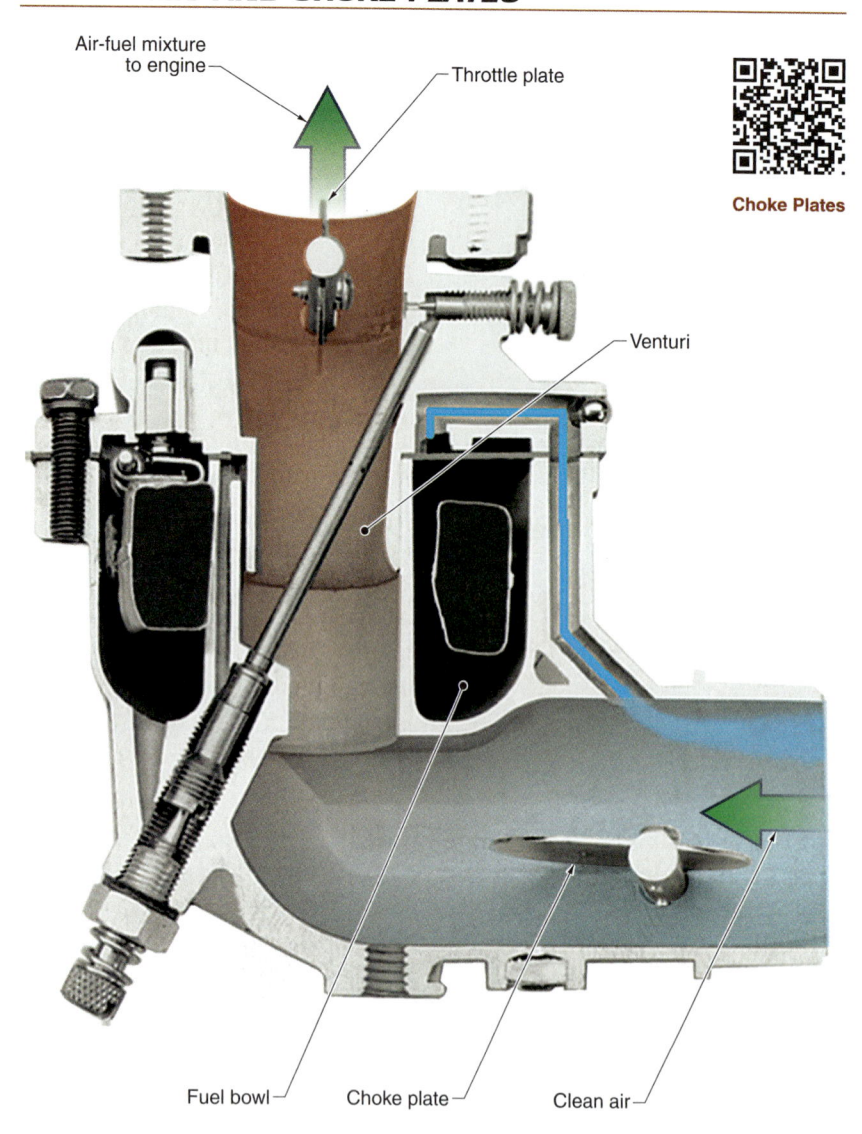

Figure 4-16. A throttle plate is used to regulate the flow of the air-fuel mixture to an engine, while a choke plate is used only to assist in cold starting.

Ignition Systems

An ignition system is the starting system for a small engine. Whether the engine is started by pulling a rewind rope or turning a key (switch) on an electric starter motor, the ignition system produces a precisely timed spark inside the combustion chamber. The ignition system includes magnets mounted on the surface of the flywheel, an ignition armature mounted adjacent to the flywheel containing copper wire windings, a spark plug lead attached to the armature, and a spark plug. *See Figure 4-17.*

When the rewind rope is pulled, the flywheel is rotated. With each rotation, the magnets pass the ignition armature. This induces electrical current flow that produces a high-voltage spark at the tip of the spark plug. The ignition system is coordinated with the piston and the valves by a flyheel key. Therefore, the spark ignites the air-fuel mixture in the combustion chamber just as the piston reaches the point of maximum compression in each engine cycle. Once the engine is running, the inertia of the flywheel keeps the crankshaft spinning until the next power stroke of the piston, while the flywheel magnets induce voltage in the armature to keep the spark plug firing.

Breaker Point Ignition Systems. A *breaker point* is an ignition system component that has two points (contact surfaces) that together function as a mechanical switch. The contact surfaces are plated with nickel. Breaker point ignition systems were used on small engines until the mid-1980s.

A breaker point ignition system functions similarly to a solid-state ignition. However, a breaker point ignition system uses a mechanical switch, rather than a transistor, to close the electrical circuit required to produce a high-voltage spark at the spark plug tip. The breaker points remain separated for most of the four-stroke cycle. A flat spot machined into the crankshaft causes one of the breaker points to pivot temporarily, closing the gap between the two and closing the circuit. The other point is retained by a spring and is mounted on a pivot. The points are held open by a point plunger. *See Figure 4-18.*

IGNITION SYSTEM COMPONENTS

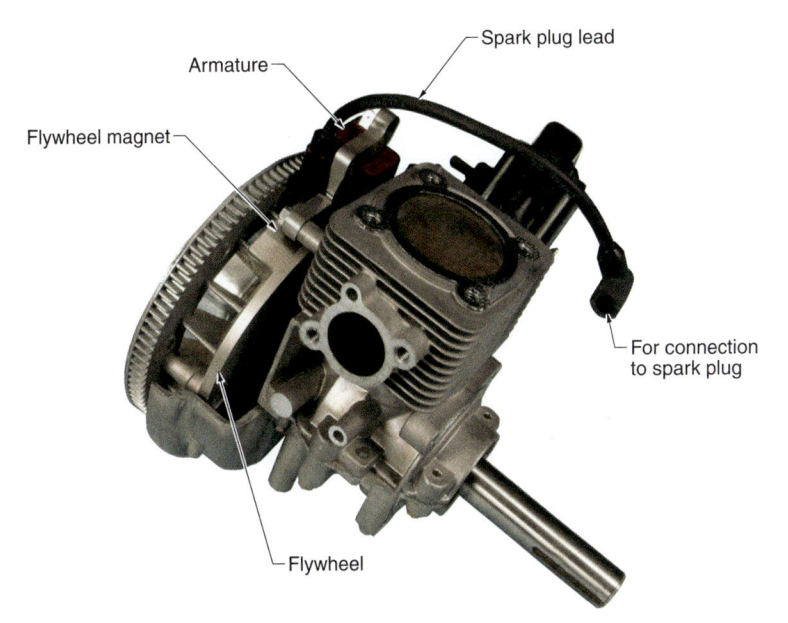

Figure 4-17. An ignition system includes multiple magnets, an ignition armature, a spark plug lead, and a spark plug.

Solid-State Ignition Systems. A solid-state ignition system requires 10,000 V to 20,000 V of electric pressure to produce a spark at the tip of a spark plug. To produce the required electric pressure in modern outdoor power equipment, a solid-state transistor is used in the ignition armature. During each pass of the flywheel magnets with the solid-state transistor laminations, the transistor establishes an electrical circuit (also referred to as closing a circuit). The circuit produces 2 A to 3 A of current. The current is then converted to high-voltage power that travels through the spark plug lead to the spark plug. *See Figure 4-19.*

BREAKER POINT IGNITION SYSTEMS

Figure 4-18. Breaker points control the flow of electricity to other parts of the ignition system circuit.

SOLID-STATE IGNITION SYSTEMS

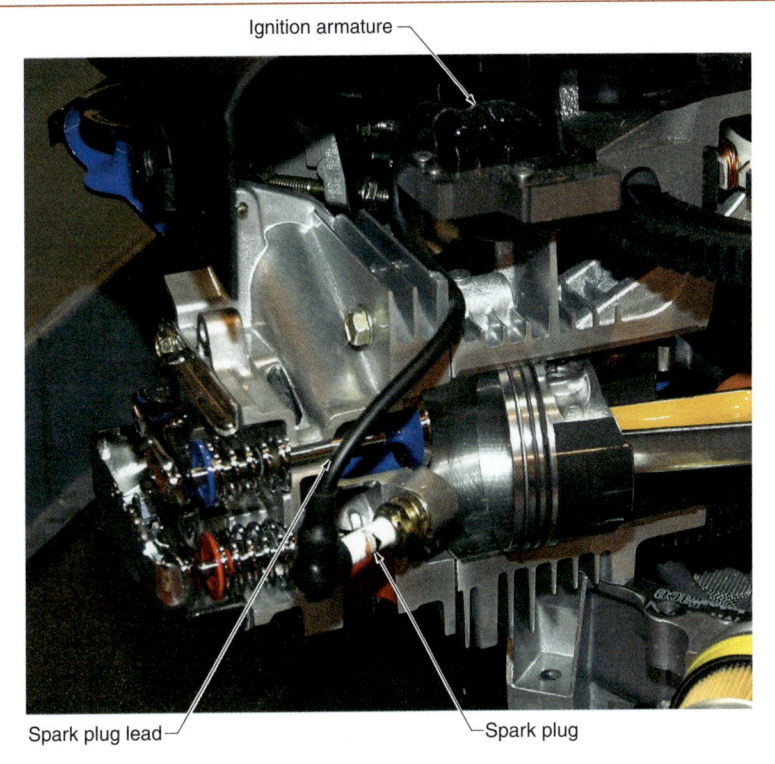

Figure 4-19. A solid-state ignition system has a solid-state transistor in the ignition armature.

Chapter 4—Small Engine Fundamentals **55**

Lubrication and Cooling Systems

The lubrication and cooling systems on small air-cooled engines are linked together. Although air passing over the aluminum cylinder, crankcase, and cylinder head carries a portion of the combustion heat energy away from the engine, the lubrication system transfers a significant portion of that heat.

Most small engines have an oil dipper that provides the motion to distribute engine oil. The lubrication system and engine oil absorb heat energy and transfer it to an aluminum cylinder block. The cooling air around the cylinder block then absorbs the heat energy from the aluminum and transfers it into the atmosphere. *See Figure 4-20.*

While the lubrication system helps to cool the engine, the engine oil maintains a working clearance between moving metal parts and bearing surfaces such as the cylinder wall and piston rings. When the system is working properly, the oil film does not allow the moving parts to touch.

LUBRICATION AND COOLING SYSTEMS

LUBRICATION SYSTEMS

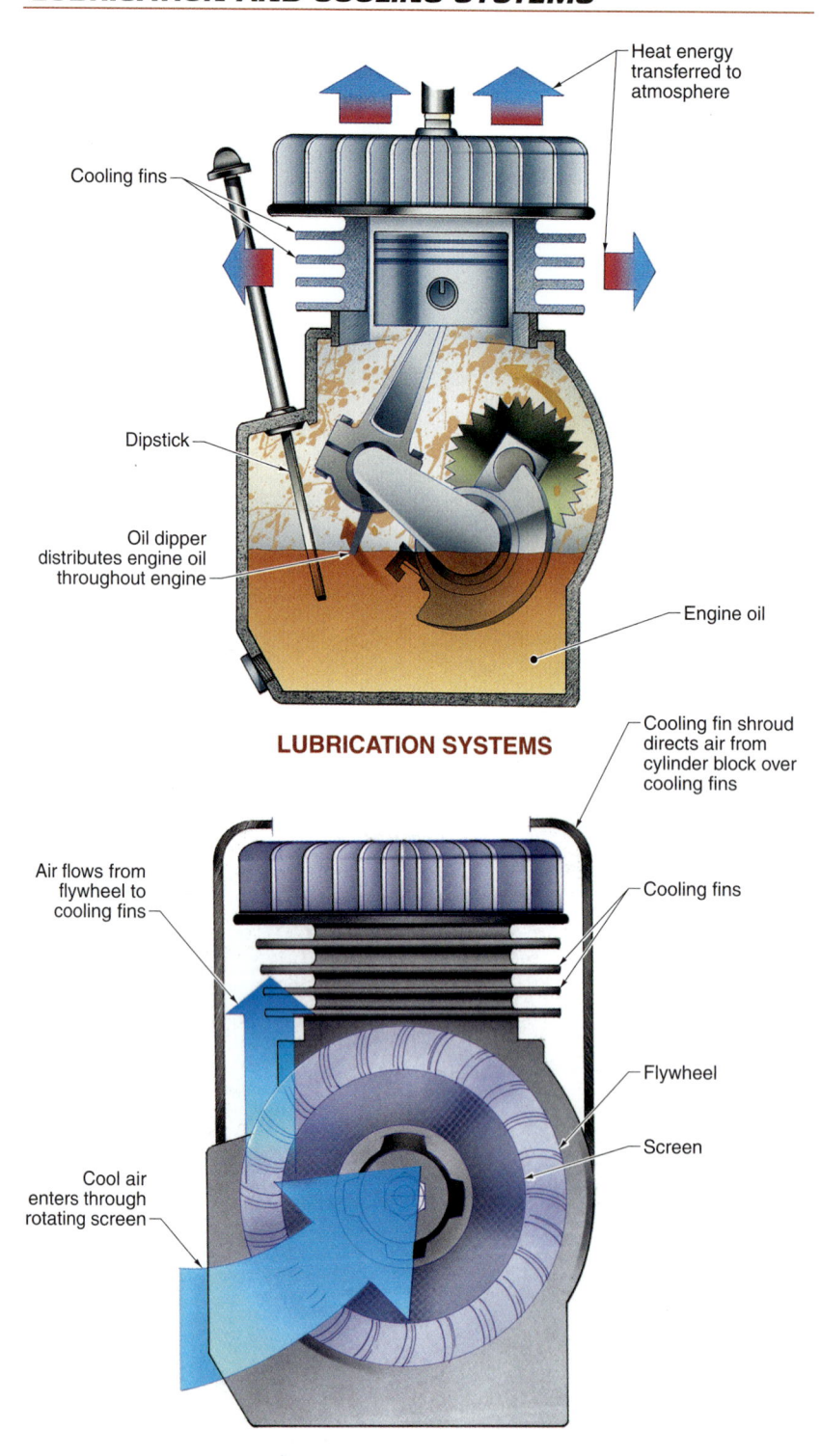

COOLING SYSTEMS (AIR-COOLED)

Figure 4-20. The lubrication system and its engine oil absorb heat energy and transfer it to the aluminum cylinder block. The cooling air can then absorb the heat energy from the aluminum and transfer it into the atmosphere.

Governor Systems

A *governor system* is an engine system that maintains a desired engine speed regardless of the load applied to the engine. ***See Figure 4-21.*** A governor system is comparable to the cruise control system of an automobile. It keeps the engine running at the selected speed, regardless of changes in the load.

GOVERNOR SYSTEMS

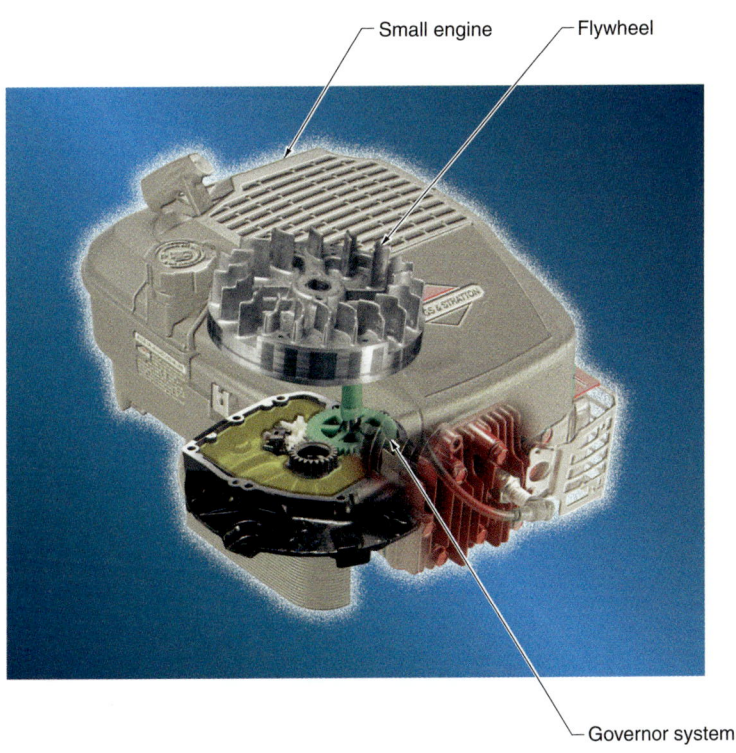

Figure 4-21. A governor system is an engine system that maintains a desired engine speed regardless of the load applied to the engine.

running lawn mower is moved from a driveway to a lawn, the crankshaft revolutions slow. However, the governor spring continues to pull, causing the throttle plate to open.

In response, a larger volume of air-fuel mixture enters the carburetor, increasing engine speed to compensate for the increased load. The crankshaft speed increases, and the pulling motion resumes until a new equilibrium is achieved. With each change in load, the tension between the governor spring and the load establishes a new equilibrium, known as the governed speed. The pulling of the springs continues until the engine is OFF. At that point, without the crankshaft spinning, the governor spring pulls the throttle to the wide-open position. The two types of governors commonly installed on small engines are mechanical and pneumatic governors.

A load is the amount of work an engine must perform to complete a job. For example, with a lawn mower, the height and thickness of the grass determines how much work the engine must perform to cut the grass to the desired height. Without a governor, the throttle on the engine would need to be manually adjusted with each pass the lawn mower makes over a dense patch of grass. A governor automatically adjusts the engine speed by detecting changes in engine speed caused by the application of a load by responding by opening or closing the throttle plate as needed to maintain the desired speed(within the limits of the engine).

The governor system operates with an unending pulling motion between at least one governor spring and a speed sensing component commonly indirectly driven by the crankshaft. The governor spring pulls the throttle toward the open position. The speed sensing device attempts to make the engine run as slow as possible. When the load on the engine increases, such as when a

Mechanical Governors. A *mechanical governor* is an engine governor that adjusts the throttle plate position as needed by using the gears and flyweights inside a crankcase as speed-sensing devices to detect changes in a load. *See Figure 4-22.* When operating a small engine under a light load, the carburetor must deliver a relatively small amount of air-fuel mixture to the combustion chamber. As the crankshaft rotates, centrifugal force causes the flyweights to open.

As they open, the flyweights apply pressure to the governor cup and governor crank, which are linked to the throttle. The throttle is pulled toward the closed position. As the load on the engine increases, the flyweights rotate more slowly. The reduced centrifugal force on the flyweights results in less pull on the throttle toward the closed position. Since the governor spring tension remains constant based on the throttle control setting, the throttle reopens until the desired governed speed is achieved.

Pneumatic Governors. A *pneumatic governor* is an engine governor that uses a movable metal or plastic air vane as a speed-sensing device to detect changes in a load. It accomplishes this by registering the change in air pressure around the rotating flywheel. *See Figure 4-23.* The design of a pneumatic governor is simpler than that of a mechanical governor, and its parts are easier to access. However, the design of the pneumatic governor is slightly less precise, since small particles of debris can interfere with its operation.

MECHANICAL GOVERNORS

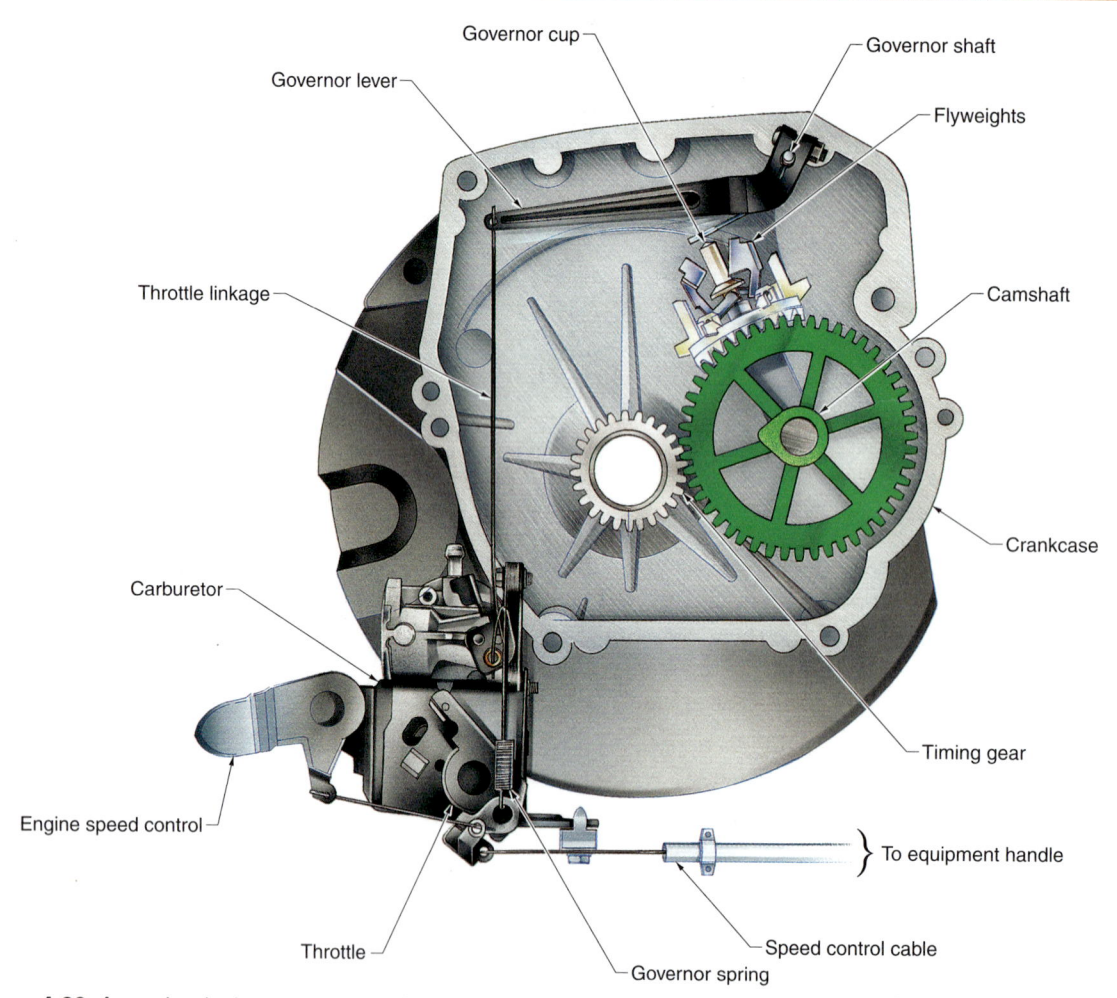

Figure 4-22. A mechanical governor uses the gears and flyweights inside a crankcase as speed-sensing devices to detect changes in a load and adjusts the throttle accordingly.

PNEUMATIC GOVERNORS

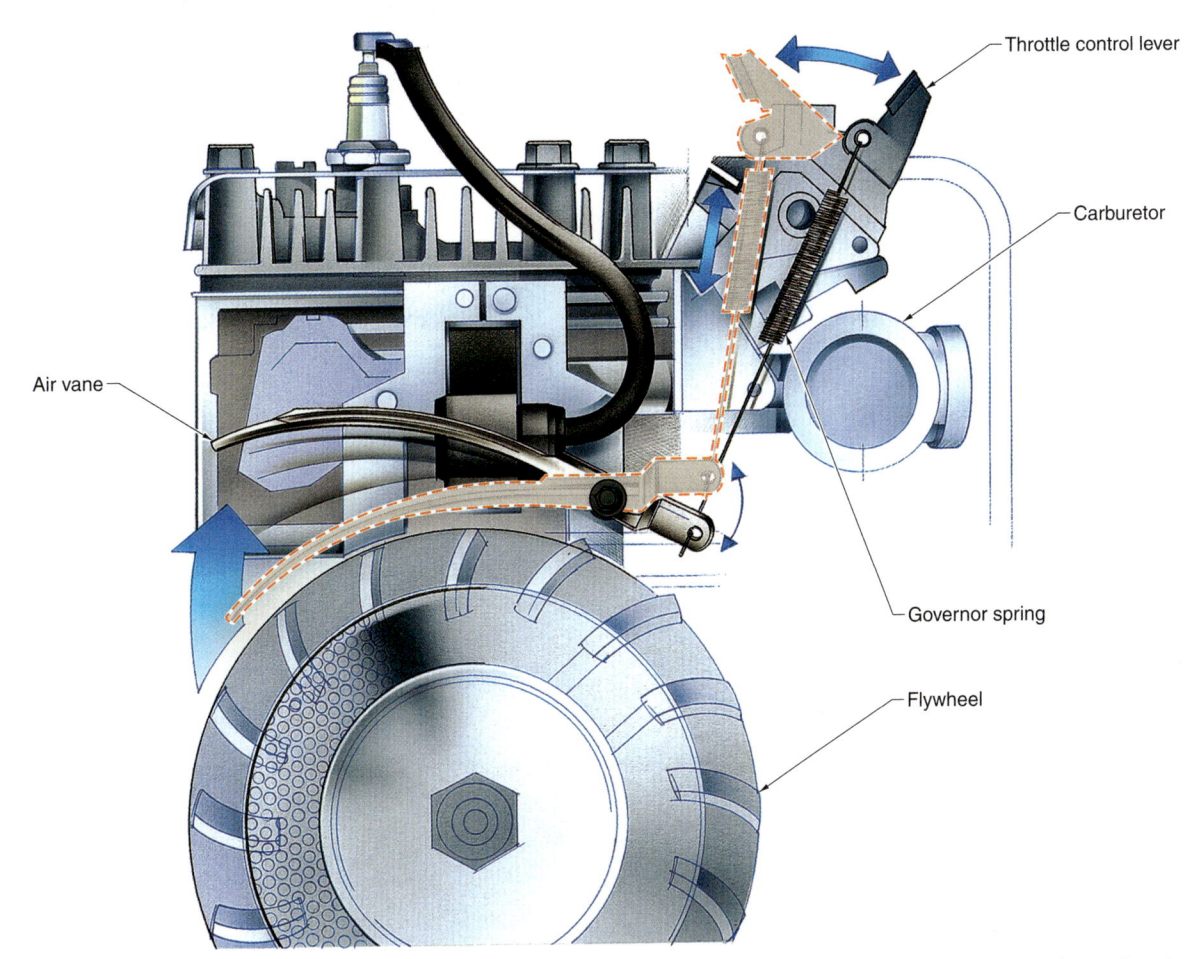

Figure 4-23. A pneumatic governor uses a movable metal or plastic air vane as a speed-sensing device by registering the change in air pressure around a rotating flywheel.

A pneumatic governor also relies on one or two springs to pull the throttle toward the open position. As the load lessens and engine speed increases, airflow created by the rotating flywheel also begins to increase. This increase in airflow causes the governor blade to pull the throttle plate toward the closed position in its effort to maintain a steady engine speed.

Electrical Systems

Batteries are installed on some outdoor power equipment, similar to automotive batteries, to operate components such as headlights, motors, and engine starters. When an engine uses the energy stored in a battery for starting, it relies on an electrical system to maintain battery strength. The electrical system of a small engine typically consists of an alternator, rectifier, regulator, and 12 V battery. *See Figure 4-24.*

An *alternator* is a charging system device that produces alternating current. An alternator consists of an assembly of at least one copper winding (stator) and a set of magnets. A *stator* is an electrical component that has a continuous copper wire (stator winding) wound on separate stubs exposing the wire to a magnetic field.

As with the ignition system, the flywheel creates a moving magnetic field to induce current. Most stators consist of a band of nonadjustable windings mounted under the flywheel and a set of magnets cemented to the inside surface of the flywheel. On certain types of small engines, the stator consists of an adjustable armature mounted under the flywheel that relies on the same magnets as the ignition armature to charge the battery. This results in longer periods of time between voltage and current surges. Limited amounts of DC voltage and current are produced, and a capacitor is commonly used to handle fluctuations in the voltage output.

Alternator Function

Alternating and Direct Current. An electrical system can be designed to produce either alternating current (AC) or direct current (DC). *Alternating current (AC)* is current flow that reverses direction at regular intervals. *Direct current (DC)* is current that flows in one direction only. See Figure 4-25.

ELECTRICAL SYSTEMS

CURRENT TYPES

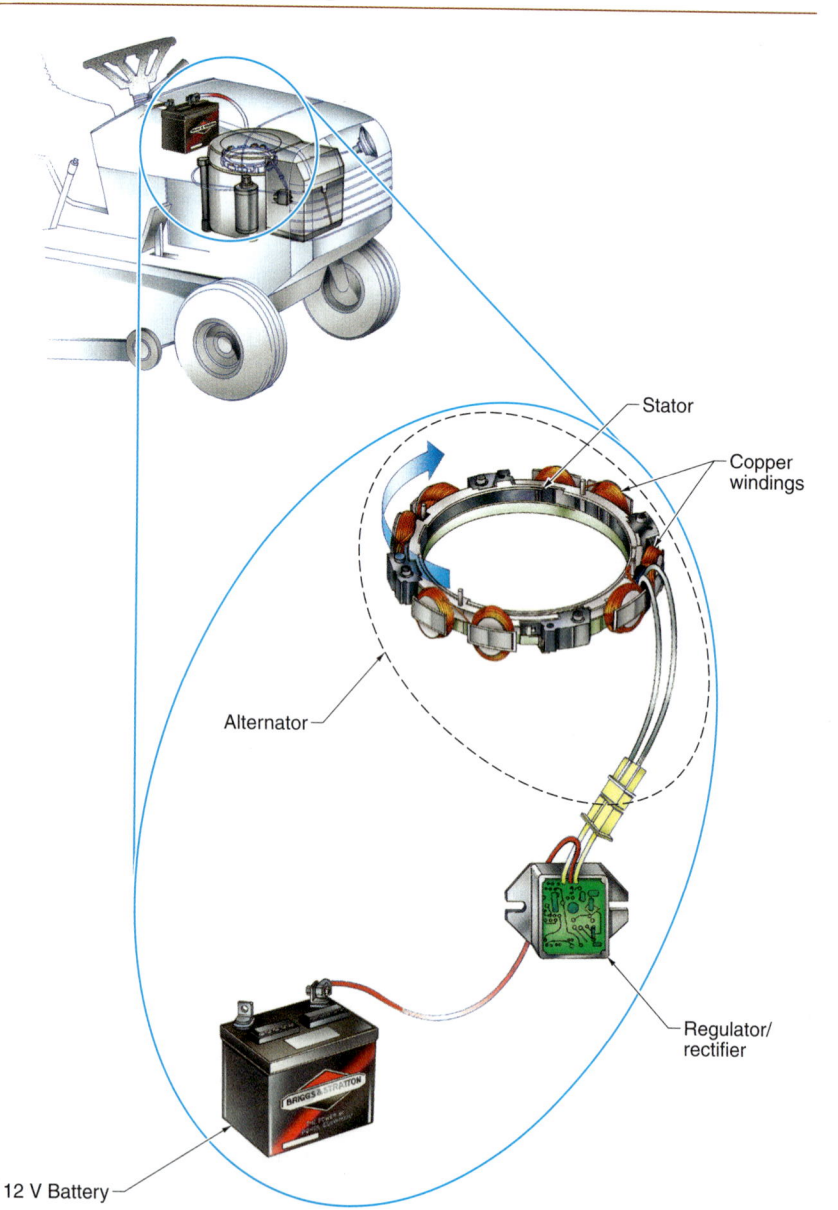

Figure 4-24. A small engine electrical system typically consists of an alternator, rectifier, regulator, and 12 V battery.

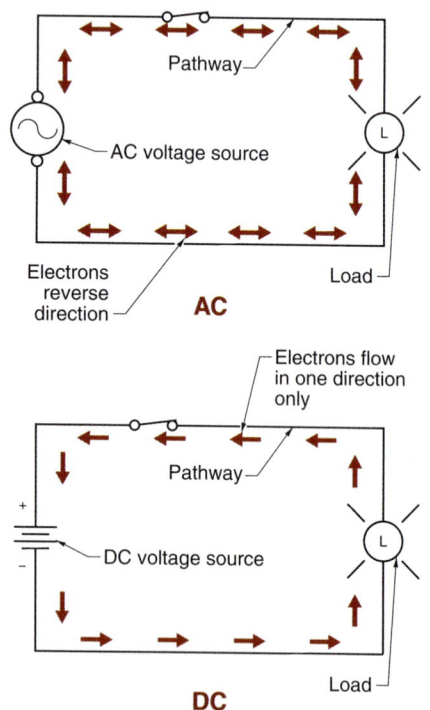

Figure 4-25. An electrical system can be set up to produce either alternating current (AC) or direct current (DC).

TECH FACT

Materials that conduct electricity are metals such as copper and aluminum. Materials that do not conduct electricity are insulators such as plastic, rubber, paper, and air.

For example, if a piece of outdoor power equipment operates lights and no other electrical component, the alternator operates like the generator on a bicycle wheel. It keeps the lights running with AC as long as the bike wheel (or the engine crankshaft) is rotating. If the outdoor power equipment includes a battery and various electrical devices, a rectifier is attached to the alternator to convert AC power to DC so that it can be stored in the battery.

DC power can be used to operate lights even when the engine is OFF, as well as a starter motor, electric cutting blade clutch, winch, and other devices that require electric power. Some engines supply AC for lights and DC for other devices. Engines that operate at high speeds also require a regulator or a combined regulator/rectifier to maintain a steady voltage output.

Braking Systems

The braking system of a modern small engine protects the user and others in the area from the moving parts of unattended equipment. The system is designed to stop the engine any time the controls are released or the operator exits the equipment such as a ride-on lawn mower. Small engine braking systems include a stop switch and stop switch wire, brake bracket, brake pad, brake spring, brake anchor, cable, operator presence control device, and throttle stop switch. *See Figure 4-26.*

On many small engines, a brake applies pressure to the smooth outer surface on the flywheel. The surface area of the brake varies in size, depending on the equipment. Some models use a brake pad while others use a brake band, which is applied to a larger area on the surface of the flywheel. Both are highly effective when properly maintained.

Most engines contain one or more stop switches wired between the ignition system and engine and equipment components. The stop switch is activated by releasing a operator presence control device or by lifting off of the seat of ride-on equipment. It interrupts power to the engine by grounding one of the copper windings in the ignition armature. When the operator presence control device is released, a wire attached to the armature is grounded against a metal engine part, stopping the engine. If the engine is equipped with a brake, the brake and brake pad, or band, apply pressure simultaneously to the flywheel. When properly maintained, the components of the braking system stop the engine within 3 sec.

BRAKING SYSTEMS

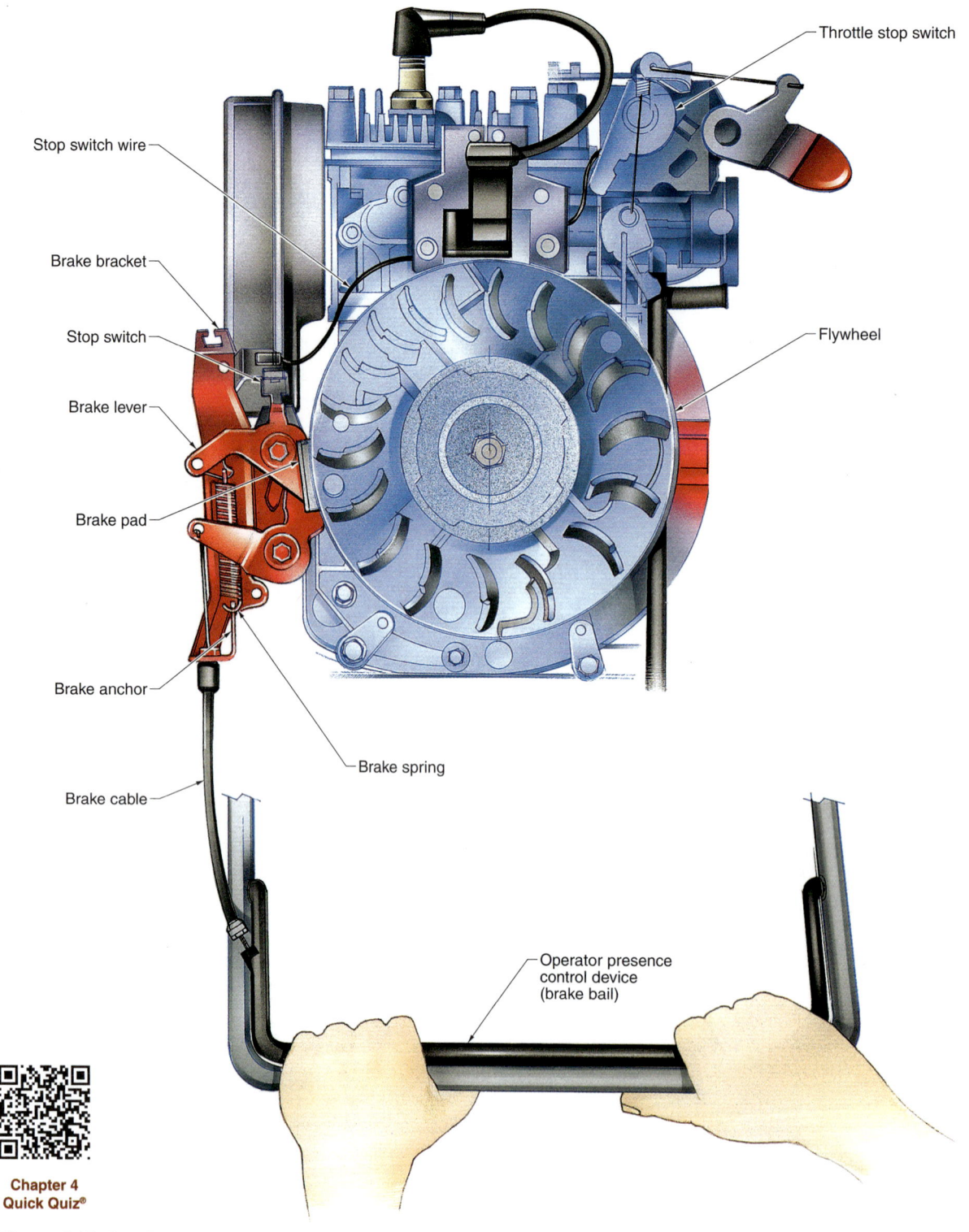

Figure 4-26. Small engine braking systems include several components that are designed to stop an engine quickly.

Outdoor Power Equipment Devices 5

INTRODUCTION

Many different devices are used to operate outdoor power equipment (OPE). The main types of these devices include mechanical drive systems and mechanical switches. Various tasks can be performed using OPE. However, the most common tasks performed are cutting grass and removing snow. Different implements and attachments can be used also with OPE to easily perform work. To ensure maximum performance and operational life, OPE must be properly maintained and stored when not in use.

MECHANICAL DRIVE SYSTEMS

A *mechanical drive system* is a drive system used to transfer power from one location to another, typically from an engine to an implement or equipment drive train. Most mechanical drive systems are designed to alter the intensity, direction, and speed of an applied driving or driven force. For example, multiple drive systems are commonly used to propel lawn tractors across a lawn and enable them to cut grass. Regardless of the method used, simple drive systems provide power to most outdoor power equipment (OPE) used for home, work, or recreational applications. The mechanical drive systems commonly used on OPE include hydrostatic drives, disc drives, and flexible belt drives.

Hydrostatic Drives

A *hydrostatic drive* is a power transmission device that uses pressurized fluid to provide power to equipment without direct contact between driving and driven components. A hydrostatic drive is powered by the engine to pressurize fluid (hydrostatic transmission fluid) in the system. Pressurized fluid is used to transfer motion to the driven side of the system in a controllable, predictable manner. The use of pressurized fluid commonly results in smooth transitions during speed changes.

There are many different designs of hydrostatic transmissions. However, all hydrostatic transmissions have components that enable them to transmit engine power to the wheels of the product, resulting in user-controlled speed and directional settings. Hydrostatic transmissions are known for their reliability, smoothness of operation, and low maintenance requirements. They are commonly installed on midgrade and premium equipment. ***See Figure 5-1.***

HYDROSTATIC DRIVES

Figure 5-1. A hydrostatic drive is a power transmission device that uses pressurized fluid to provide power to equipment without direct contact between driving and driven components.

Disc Drives

A *disc drive* is a power transmission device that uses a friction disc to make contact between the driving and driven components of equipment. Typically, a disc drive has a rubberized polymer molded to the circumference of the disc. The rubberized polymer (friction) surface is placed in contact with a drive disc, which is connected to the engine. The position of the friction disc in relation to the driven disc, through the use of tension springs, results in variable speed control. Variable speed control results in a smooth transition of gears while changing speeds and a simplified method of power delivery to the application. Also, the position of the driven disc in relation to the drive disc determines the speed of the driven disc. **See Figure 5-2.**

Flexible Belt Drives

A *flexible belt drive* is a mechanical drive system that uses a flexible belt to transfer power between a drive and driven shaft. A flexible belt drive system is the most commonly used mechanical drive system for OPE because it is inexpensive, quiet, and easy to maintain. A V belt is the most commonly used type of flexible belt in the OPE industry. Proper belt alignment and tension are critical to the operation of any flexible belt drive system.

A *V belt* is a reinforced rubberized belt that connects pulleys to rotate components. The cross section of the V belt is in the shape of a "V". The belt fits precisely into a V-shaped pulley on either the drive side or driven side of the flexible belt system. **See Figure 5-3.** The friction between the surface of the belt and the surface of the pulley results in the transfer of rotation and power. Belt tension is produced from the belt being pulled or stretched. For proper operation, it is important that the belt be tensioned correctly. A V belt must be tight enough to drive the pulley yet loose enough when the tension is released to disengage from the pulley and stop transferring rotation and power.

DISC DRIVES

Figure 5-2. A disc drive is a power transmission device that uses a friction disc to make contact between driving and driven components.

TECH FACT

Belt wrap is the percentage of the belt surface that is in contact with the pulley in a belt drive system. Belt wrap is determined by the size of the pulley and the cross section of the belt. The optimum belt wrap (100%) of a V belt and pulley is 180°. This means the belt is in contact with the pulley for one-half of the outer diameter of the pulley.

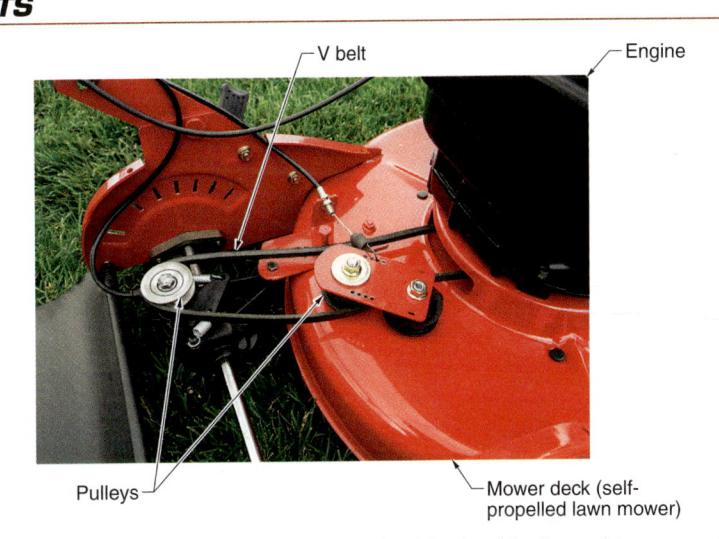

V BELTS

Figure 5-3. A V belt is a continuous reinforced rubberized belt used to connect pulleys in order to transfer rotation.

A *belt guide* is a component used with a belt drive system to retain a flexible belt within the confines of a pulley groove. A belt guide is not designed to help with the engagement of the belt but rather with disengagement. The correct placement and clearance between the belt guide and the belt are required for proper operation.

Pulleys. Many mechanical drive systems include pulleys. A *pulley* is a grooved rotating wheel with a belt around a portion of its circumference that is used for transferring power. Pulleys are used with OPE to transfer power through the use of a drive belt. Pulleys are made from various metals or plastics. The types of pulleys used for OPE include drive, driven, and idler pulleys. The most common pulley configuration used for OPE is the V-shaped pulley. *See Figure 5-4.*

PULLEYS

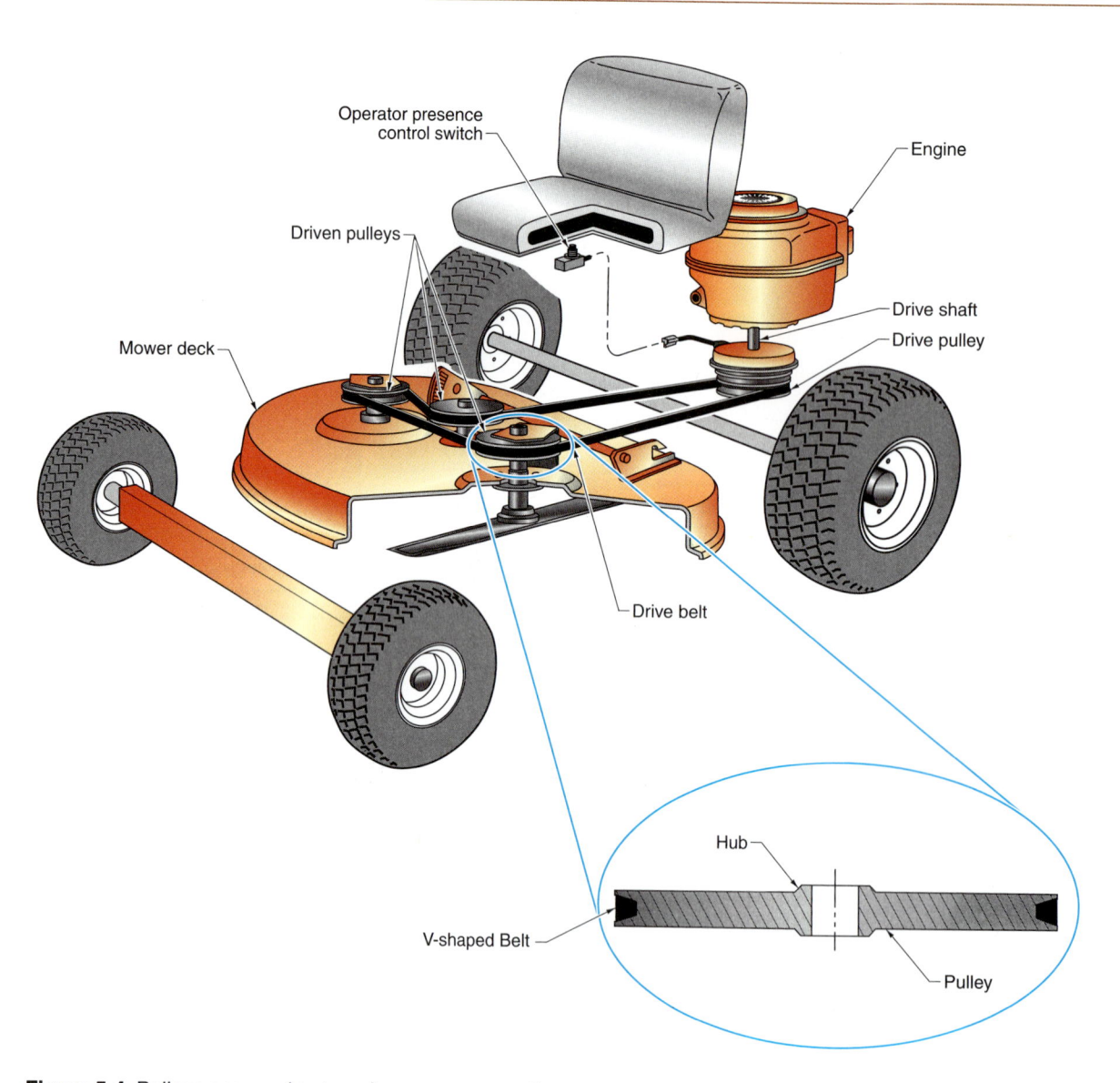

Figure 5-4. Pulleys are used to transfer power among the components in equipment through the use of a drive belt.

MECHANICAL SWITCHES

A *mechanical switch* is an electrical control device used to allow, interrupt, or direct the flow of electricity through a circuit. Small engines use mechanical switches to allow engines to be started with an electric starter motor circuit or allow the engine to be stopped through the use of an ignition circuit. Mechanical switches allow the use of various safety circuits and attachment power circuits in OPE. Manufacturers install the various types of mechanical switches in equipment based on the intended application, switch design requirements, and circuit demands. The most commonly used types of mechanical switches are safety interlock switches and power takeoff switches.

Safety Interlock Switches

A *safety interlock switch* is an electrical control device that helps prevent the unsafe operation of equipment. Broadly used in the industry, safety interlock switches generally can be divided into two categories: operator presence control switches and ignition switches. **See Figure 5-5.**

SAFETY INTERLOCK SWITCHES

Switch exposed under raised operator seat

OPERATOR PRESENCE CONTROL

Switch visible on console

IGNITION

Figure 5-5. The most commonly used safety interlock switches are operator presence control switches and ignition switches.

Operator Presence Control Switches. An *operator presence control switch* is a switch used to ensure the safe positioning of an equipment operator relative to the equipment. For example, a seat switch, installed under the seat of ride-on OPE, is connected in such a way as to not allow the engine to be started or continue to run under the following circumstances:

- the operator is not in the sitting position
- the operator does not have proper hand control
- the operator is not in a safe operating position for equipment use
- the operator does not have the control to stop an attachment from operating when the transmission is put into reverse

Ignition Switches. An *ignition switch* is a rotary-actuated, double-pole double-throw (DPDT) switch that uses a key as an actuating mechanism. The ignition switches used on small engines and OPE are unique in operation and appearance. The ignition switch usually mounts in a centrally located position as part of a dashboard. The side of the switch that faces the operator has a hole for the ignition key.

The opposite side of the switch has three to six electrical connection spades where circuits are connected. Although there are exceptions, most OPE ignition switches have five or six electrical connection spades.

The key acts as the actuating mechanism and allows the operator to turn the switch to various positions. The switch positions are described by the related positions of the hands on a clock. For example, the 12 o'clock position is the OFF position and is achieved by rotating the key counterclockwise to its farthest position. Rotating the key clockwise to the next position places the key at the 2 o'clock position, which is the run position. The 4 o'clock position is the start position. The 6 o'clock position of an ignition switch is an intermittent position and spring-loaded internal connection. When released from the start position, the key rotates automatically under spring tension back to the run position.

Operator Presence Control Switches

Power Takeoff Switches

A *power takeoff (PTO) switch* is a switch used to control the PTO function on a lawn and garden tractor. For example, a lawn and garden tractor may be equipped with an electric-powered clutch, which drives a V belt. When the PTO switch is actuated into the ON position, the electric circuit that engages the clutch is actuated. Then, the mower deck or other attached implement is engaged.

Many PTO switch circuits include a reverse mowing lockout so that the operator cannot mow in reverse without actively engaging an additional safety switch. The actuating mechanism of a PTO switch is typically a large button that controls the in-out travel movement of the switch. *See Figure 5-6.* Since PTO switches are usually designed to handle large current loads, some circuit designers allow the operating current for the clutch to run directly through the switch.

POWER TAKEOFF (PTO) SWITCHES

Figure 5-6. A PTO switch is used to control a PTO function on a lawn and garden tractor. It is typically located on the control panel of OPE.

Accessory Switches

An accessory switch is used to control auxiliary equipment. Mechanical accessory switches are used in headlights, audible alarms, low oil warnings, and hour meters. Although many accessory electrical circuits can be integrated into more complex switches, most accessory switches are single-pole single-throw (SPST) switches.

IMPLEMENTS AND ATTACHMENTS

Various implements and attachments can be used with OPE. An *implement* is a device connected to, and typically powered by, a machine that is used to perform a specific task. For example, snow throwers, backhoes, brooms, front-end loaders, and rotary mower decks connected to lawn tractors are considered implements. An *attachment* is a device connected to a machine or implement that is used to perform a task. Attachments perform work through the pulling or pushing action of the equipment to which they are attached such as tractors. For example, scrapers, plows, and tillers are considered attachments. ***See Figure 5-7.***

Some implements and attachments can perform many different types of tasks. However, most implements and attachments are used for either cutting grass or removing snow.

IMPLEMENTS AND ATTACHMENTS

Snow Thrower

Backhoe

Broom

Front-end Loader

Rotary Mower Deck

IMPLEMENTS

Rear Scraper Blade

Rear Plow

Plow Blade

Front Scraper Blade

Tiller

ATTACHMENTS

Figure 5-7. A variety of implements and attachments can be used with OPE.

Grass-Cutting Devices

The most common implement used with OPE is a rotary mower deck. A *rotary mower deck,* also known as a mower deck, is an implement that uses one or more flat blades to cut grass with a circular motion. A mower deck is designed for cutting grass only.

Components commonly found on a mower deck include a flexible drive belt, belt guide, idler pulley, blade bearing quill, and blade. A mower deck is typically powered by an engine through the use of a flexible belt. The flexible belt connects the power source to the implement. When the belt is engaged, rotary cutting blades rotate and slice the grass to a desired height. Typically, large mower decks have three rotary cutting blades. *See Figure 5-8.* The leading edge of the blade is sharpened at an angle, providing a cutting surface. A mower deck, or other grass-cutting implement, may have a single rotating blade or multiple rotating blades.

Single Blades. The rotating blade of a single-blade mower deck is typically 19″ to 22″ in length. The blade shape and size are determined by the intended purpose of the blade. For example, some single-blade cutting systems are designed to allow the operator the option of bagging grass clippings as the lawn is cut. With this design, the blade typically has a high lift angle on the trailing side of the blade. The high lift angle creates increased air flow, which lifts the cut grass and directs it to the bagger system. Other mowers produce mulch by chopping the grass clippings into fine particles, which is then discharged back onto the surface of the lawn. The blade produces the finer clippings.

GRASS-CUTTING DEVICES

SINGLE BLADE

Mower deck with one blade

MULTIPLE BLADES

Mower deck with three blades

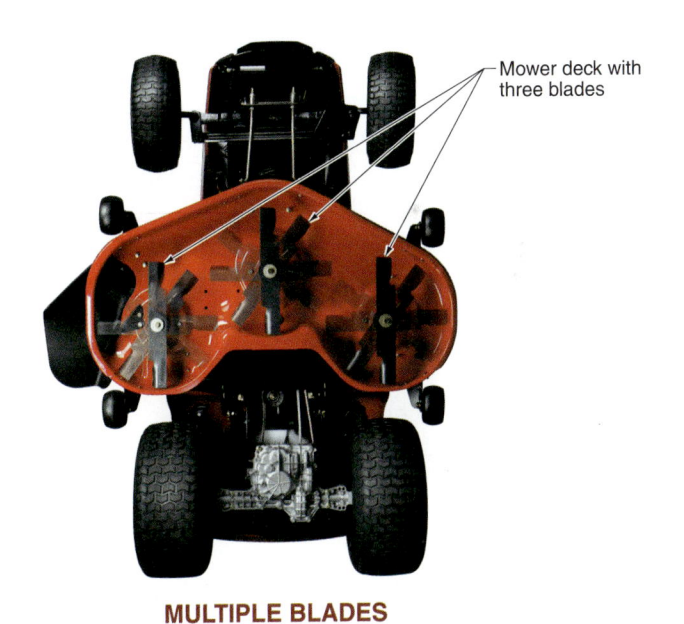

Figure 5-8. The most common implement for OPE is the single- or multiple-blade rotary mower deck.

Multiple Blades. A multiple-blade mower deck typically has two or three rotating blades. However, configurations of up to five blades are available. Most multiple-blade systems are asynchronous. An asynchronous system does not provide a constant, positive engagement between the drive and the driven member. This means there is no specific timing of the blade position during the operation of the system. A limited number of manufacturers use synchronous systems, which provide constant, positive engagement between the drive and driven members. The flexible drive belt of a synchronous system has segmented teeth, or notches, on the base of the friction surface. The belt must be properly installed and timed with the operation of the blades.

Snow-Removal Devices

In cold climates, a snow thrower is often used. A snow thrower can be attached to a larger piece of OPE, such as a lawn and garden tractor, or can be purchased as a stand-alone, single-purpose piece of OPE. *See Figure 5-9.* A snow thrower operates by gathering snow into a collection hopper and discharging the collected snow out of a chute some distance from the unit. The two types of snow thrower attachments are single-stage and two-stage snow thrower attachments.

SNOW-REMOVAL DEVICES

Two-stage snowthrower — Lawn and garden tractor

Figure 5-9. Snow throwers can be attached to a larger piece of OPE such as a lawn and garden tractor.

Single-Stage Snow Throwers. A *single-stage snow thrower* is a snow thrower that removes snow using a single rapidly rotating auger. An *auger* is a rotating shaft with curved paddles extending outward from its center. As the auger rotates at high speed, snow is gathered into the collection hopper which is attached to a discharge chute. The speed and power of the auger and the forward walking motion of the operator determine the amount of snow gathered and the distance it is discharged.

Two-Stage Snow Throwers. A *two-stage snow thrower* is a snow thrower that includes a single-stage auger and a blower fan. The *blower fan* is a rotating disc, driven by a V belt, with multiple paddles. The operation of a two-stage and single-stage snow thrower is similar, except that as snow is gathered into the collection hopper of a two-stage snow thrower, the blower fan (which rotates perpendicular to the auger) increases the speed of snow removal and amount of snow discharged.

OPE MAINTENANCE

The components of OPE that must be regularly maintained are the engine, the drive system, and the various implements and attachments. Cleaning is an integral part of regular OPE maintenance. In normal operating environments, OPE is exposed to dirt and debris, which can accumulate in and around moving parts. Since the presence of dirt and debris plays a role in component failure, neglecting to clean parts properly usually results in poor equipment performance and increased repair costs. Generally, a water hose or pressure washer can remove unwanted materials from equipment sufficiently. *See Figure 5-10.*

CLEANING OPE

Figure 5-10. A water hose or pressure washer can be used to remove unwanted materials from OPE, allowing visual access to most moving components.

Engine Maintenance

Many small engine problems can be avoided by following a regular maintenance schedule. The costs associated with part replacement and repair can also be reduced by regular engine maintenance. *See Figure 5-11.* Small engines that are used frequently or exposed to dirty and dusty environments must be serviced more frequently than those that operate under less severe conditions. Maintenance should begin when the engine is new.

ENGINE MAINTENANCE

After First Five Hours of Use
- Change oil and filter.

Before and After Each Use
- Check oil.
- Remove debris around muffler.

Every 25 Hours of Use or Every Season
- Check oil if operating under heavy load or in hot weather.
- Service air cleaner assembly.
- Clean fuel tank and line.
- Clean carburetor float bowl, if equipped.
- Inspect rewind rope for repair.
- Clean cooling fins on engine block.
- Remove debris from blower housing.
- Check engine compression.
- Inspect governor springs and linkages.
- Inspect ignition armatures and wires.
- Inspect muffler.
- Check valve tappet clearances.
- Replace spark plug.
- Adjust carburetor.
- Check engine mounting bolts/nuts.

Every 100 Hours of Use or Every Season
- Clean cooling system.
- Change oil filter, if equipped.
- Decarbonize cylinder head.

WARNING: A muffler becomes extremely hot during engine operation and can burn skin. Allow engine to cool a minimum of 30 min before performing any maintenance on or near the muffler.

Figure 5-11. Many small engine problems can be avoided by following a regular maintenance schedule, which reduces the costs associated with part replacement and repairs.

Each piece of OPE comes with an owner's/operator's manual provided by the original equipment manufacturer (OEM). The manual of a particular make and model of OPE should be consulted for regular maintenance guidelines. If a hard copy is not available, a digital copy of the manual often can be found on the website of the OEM.

Drive System Maintenance

Drive systems are divided into three basic categories: hydrostatic, flexible belt, and disc. Each category has specific maintenance requirements. For example, nearly all hydrostatic drive systems are required to be serviced by professionals. Hydrostatic drive systems do not have any user-recommended maintenance requirements. Flexible belt and disc drive systems, however, can be adequately maintained by nonprofessionals, with some acquired knowledge.

A flexible belt drive is the most elementary part of a drive system, in regard to complexity. Flexible belt drives must be kept clean and dry to operate properly. Even small amounts of moisture from rainwater or cleaning materials can disrupt the proper operation of the most well-maintained drive system.

74 SMALL ENGINE AND EQUIPMENT MAINTENANCE

The maintenance of flexible belt drives consists of belt adjustment and replacement. A flexible belt does not have an infinite life span. A belt replacement schedule is based on the number of hours the belt has been in use and the amount of time elapsed. Time, sun exposure, lubricant exposure, and wear are factors that contribute to belt deterioration and result in the increased need for maintenance. When replacing a flexible belt, it is helpful to sketch a belt routing diagram on a sheet of paper prior to removing the existing belt. This helps to ensure that the replacement belt is properly routed. *See Figure 5-12.*

Disc drive systems are most commonly used to cause the rotational movement of equipment such as augers. Because system operating pressures and wear of a propulsion system vary based on load, hours of use, and terrain, the required frequency of disc drive maintenance varies. The maintenance of a disc drive system usually involves unbolting several components, installing a new disc, and adjusting components. *See Figure 5-13.* Since the system is relatively simple in design, the maintenance for this type of drive system can be performed by a moderately skilled nonprofessional.

Implement and Attachment Maintenance

Proper maintenance extends the life of implements and attachments. Because implements and attachments, such as mower decks and snow throwers, are seasonal devices, they are not usually connected to OPE and are kept in storage until needed. In order to operate properly when needed, these devices must be properly maintained.

FLEXIBLE BELT REPLACEMENT

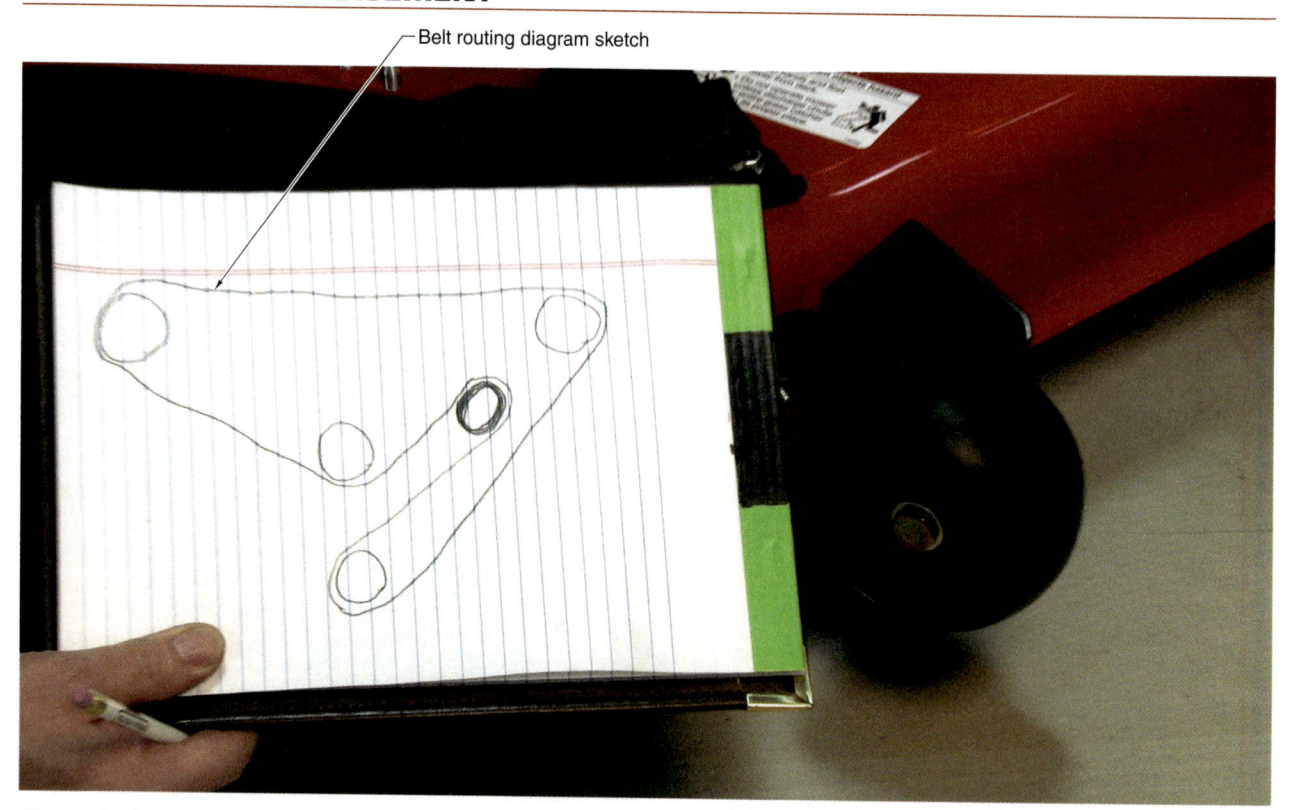

Figure 5-12. When replacing a flexible belt, it is helpful to sketch the belt routing diagram on a sheet of paper prior to removing the existing belt.

DISC DRIVE REPLACEMENT

Figure 5-13. The maintenance of a disc drive system usually involves unbolting several components and installing a new disc followed by an adjustment procedure.

Mower Deck Maintenance. The proper maintenance of rotary mower decks includes the adjustment, alignment, and replacement of the flexible belt and the sharpening and replacement of the cutting blade. *See Figure 5-14.* A static, engaged flexible belt must have some deflection to ensure that the belt will disengage as required to transfer power from the engine to the implement. Proper flexible belt adjustment is required for component performance and longevity.

TECH FACT

All small gasoline engines produce toxic gases and must be operated in well-ventilated areas. Improper exhaust system modifications can result in a safety hazard. Engine installations should discharge exhaust gases away from the application and operator safely and efficiently.

MOWER DECK MAINTENANCE

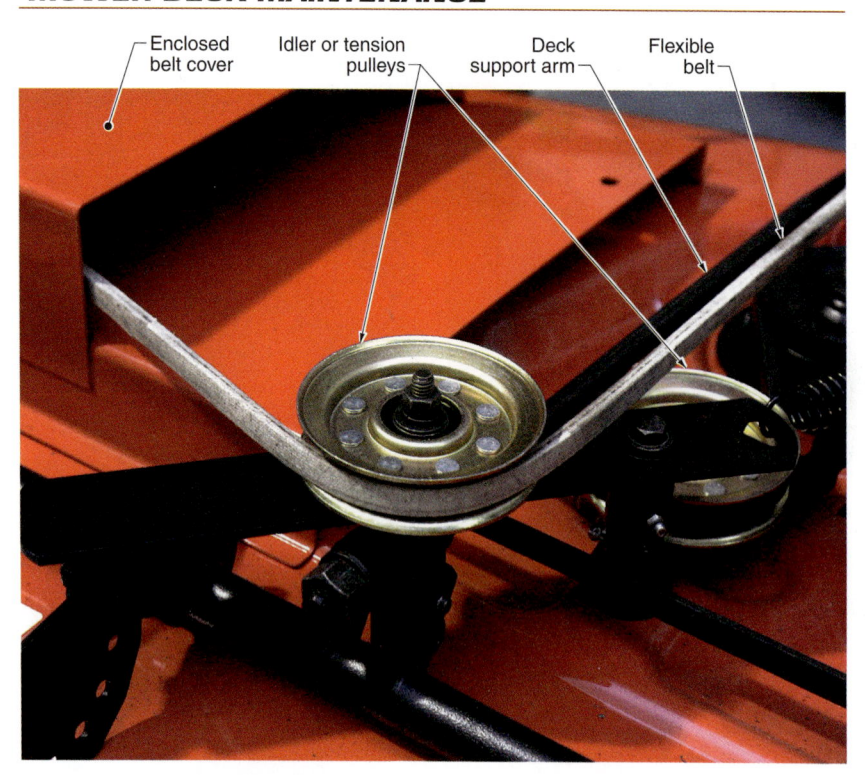

FLEXIBLE BELTS, GUIDES, AND PULLEYS

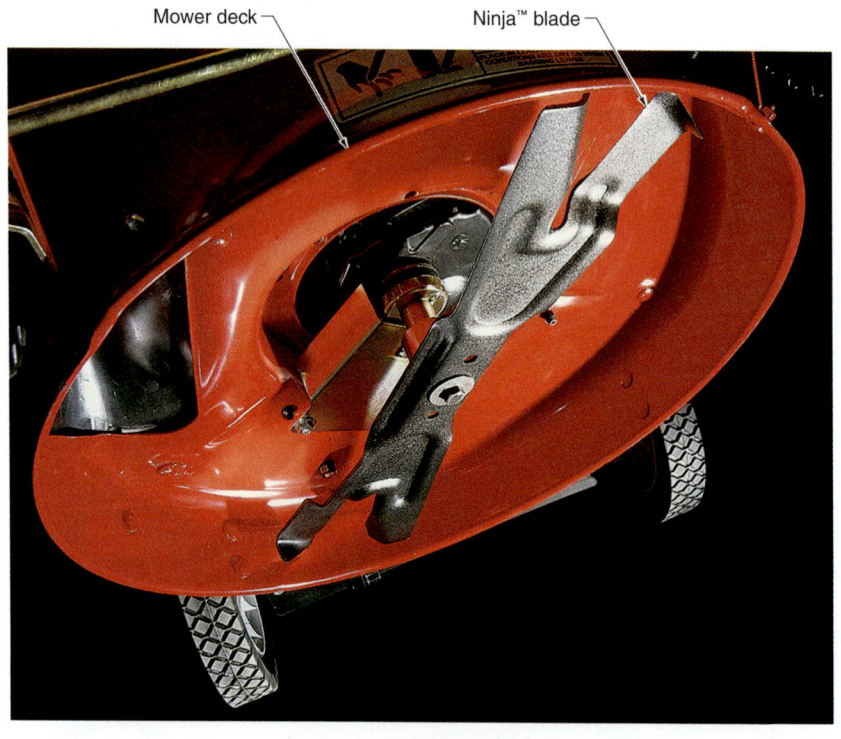

CUTTING BLADES

Figure 5-14. Mower deck maintenance consists of flexible belt adjustment and alignment, flexible belt replacement, and cutting blade sharpening or replacement.

In addition, the alignment of the blade pulleys, idler pulleys, and belt guides is important for the proper operation of the drive system. Most mower deck designs include belt guides to ensure that the flexible belt remains within the intended groove under engagement and releases from the groove when disengaged. The effects of misalignment include excessive belt wear and belt release from the pulley grooves when disengaged.

The decision to sharpen or replace a blade depends on how much of the blade has been consumed through previous sharpening and the general condition of the blade. Any blade that exhibits loss of alignment or has significant nicks should be replaced. Blades must be sharpened through a professional sharpening service or at a repair shop for lawn and garden equipment.

Snow Thrower Maintenance. A common attachment for a lawn and garden tractor is a snow thrower. The snow thrower is typically connected to the front of the tractor using a series of nuts and bolts or locator dowels and hitch pins. *See Figure 5-15.* A *locator dowel* is a cylindrical metal rod inserted into the holes in the adjacent members of a joint to align and strengthen the joint. A *hitch pin* is a pin used to secure a locator dowel when connecting two pieces of equipment. Power for the attachment is typically supplied from a flexible belt, which connects directly to a drive shaft connected to the engine.

SNOW THROWER MAINTENANCE

Figure 5-15. To properly maintain a snow thrower, it is critical that the recommendations of the OEM are followed with precision.

With some designs, there is a direct shaft connection between the attachment and the engine. Since the flexible belt can be negatively impacted by any misalignment or incorrect belt adjustment, it is critical that the recommendations of the OEM are followed with precision. Many of the maintenance concerns for mower decks, such as belt alignment, belt guide adjustment, and belt location, also apply to snow thrower attachments.

STORAGE

The proper storage of OPE is a part of preventative maintenance. Although equipment inactivity may appear to be benign, improper storage can greatly reduce the longevity and performance of any OPE. Improper storage procedures can lead to problems such as the loss of oil and fuel, the oxidation of metal parts, dry and brittle seals, and the loss of belt flexibility. Specific storage procedures are used for engines, equipment, and batteries.

Engine Storage

Storing an internal combustion engine from season to season is considered to be long-term storage by industry standards. Improper storage procedures can cause unseen deterioration of various metal, plastic, and rubberized components of the engine. Engine component deterioration can be caused by fuel degradation and exposure to extreme temperatures, dirt, and debris.

Proper engine storage significantly reduces the occurrence of seasonal start-up and performance problems associated with OPE. *See Figure 5-16.* Proper engine storage commonly includes the following tasks:

- emptying the engine of fuel
- adding a fuel stabilizer to an engine that contains fuel
- sealing the fuel tank to prevent dirt and debris from entering
- changing engine oil (or adding new oil)
- covering the engine with a plastic sheet to prevent the accumulation of dust, dirt, and other debris

A *fuel stabilizer* is a compound used to extend the life of fuel that is not or cannot be stored properly. A fuel stabilizer is typically used with small engines to promote quicker and more reliable starting. Fuel containers and tanks should be kept more than half full and properly capped to reduce air exposure. Prior to long-term storage, an engine should be run for 10 min to circulate the fuel stabilizer through all components and to purge old, stale fuel from the carburetor and fuel line. Storage at high temperatures should be avoided.

78 SMALL ENGINE AND EQUIPMENT MAINTENANCE

ENGINE STORAGE

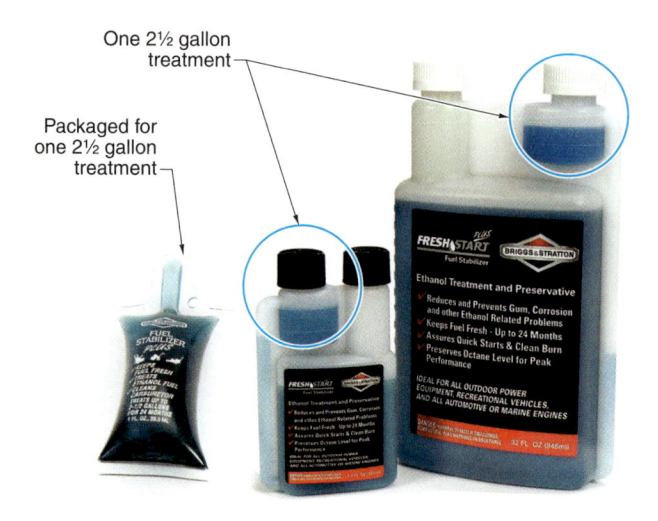

ADD FUEL STABILIZER TO FUEL TANK

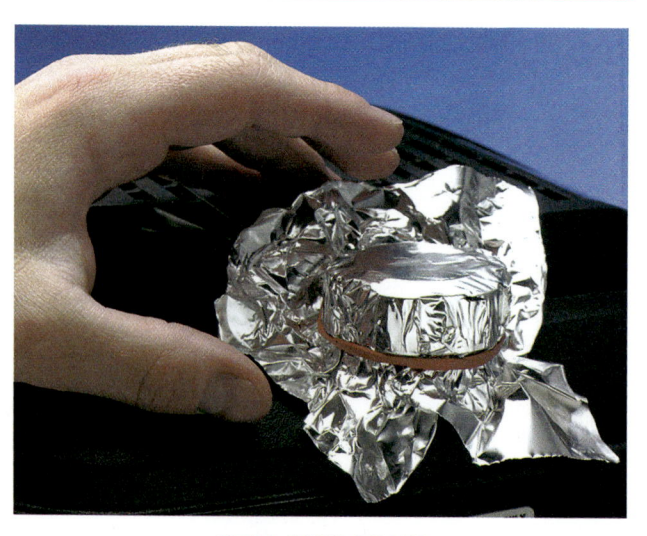

- One 2½ gallon treatment
- Packaged for one 2½ gallon treatment

SEAL FUEL TANK

CHANGE OR ADD NEW ENGINE OIL

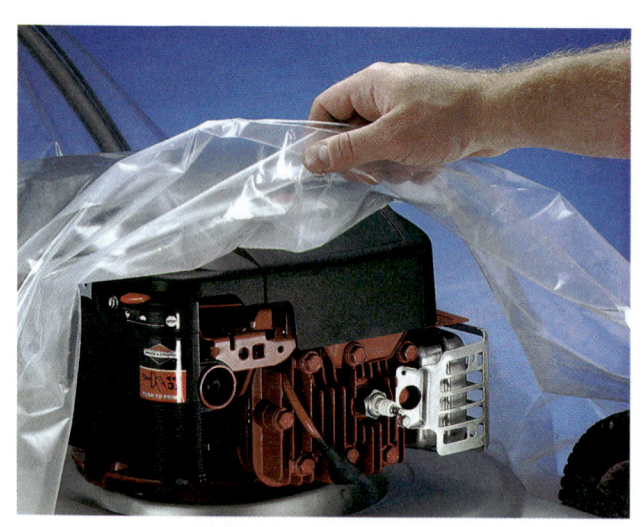

COVER WITH PLASTIC SHEET

Figure 5-16. Common engine storage procedures include adding fuel stabilizer to a tank that contains fuel, sealing the fuel tank, changing or adding new engine oil, and covering the engine with a plastic sheet.

TECH FACT

If a fuel stabilizer is not available to be added to a fuel tank prior to storing equipment, fuel should be drained from the tank or the engine should be operated until the fuel runs out.

Equipment Storage

Storage requirements specified in OEM-provided manuals may vary among OPE manufacturers. ***See Figure 5-17.*** However, there are general guidelines that must be followed and apply to all OPE. The most common equipment storage procedures require, where practical, releasing the tension of drive belts and mower deck belts to remove any lumps.

Fuel Stabilizers

OWNER'S/OPERATOR'S MANUALS

Figure 5-17. Storage requirements, which are specified in the OEM-provided owner's/operator's manual, may vary among each different piece of OPE.

Although constructed to provide above average performance in any season, long-term seasonal storage can present a challenge for even premium equipment. Many of the systems and subsystems found in OPE can become contaminated from the environment, lose their shape, or become inoperative from extreme temperature fluctuations and time. Performing simple but necessary storage procedures at the end of a season helps ensure that any type of OPE performs well after each season.

Battery Storage

Any piece of OPE that has a 12 V lead-acid battery can be stored with the battery either installed or removed. It is recommended that batteries be removed from equipment and stored in a cool, dry area to extend battery life. If, however, a battery is stored in equipment for a long period of time, regular charging (about once a month) of the battery is recommended. *See Figure 5-18.*

BATTERY CHARGING

Battery charger

12 V lead-acid battery (installed)

Figure 5-18. If a battery remains installed in equipment during long-term storage, regular charging (about once a month) of the battery is recommended.

Storage Benefits

While there are some types of OPE that can be used during any season, such as portable electric generators, pumps, log splitters, and lawn and garden tractors, there are many types of OPE that are used only during certain seasons. They must be properly stored once the season in which they are used ends. The most common types of OPE that require seasonal storage include rotary lawn mowers, snow throwers, pressure washers, string trimmers, and lawn edgers. *See Figure 5-19.* In order to prolong the service life of these pieces of OPE they must be properly stored after each season. Failure to do so can result in inoperable equipment at the beginning of the next season of use.

SEASONAL OUTDOOR POWER EQUIPMENT

ROTARY LAWN MOWERS

SNOW THROWERS

STRING TRIMMERS

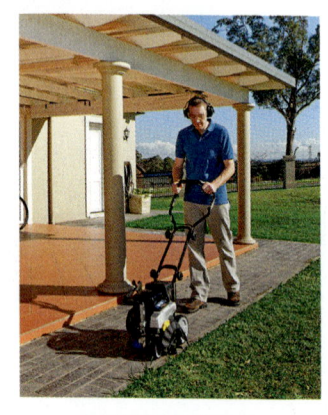

LAWN EDGERS

PRESSURE WASHERS

Figure 5-19. The most common types of OPE that require seasonal storage include rotary lawn mowers, snow throwers, string trimmers, lawn edgers, and pressure washers.

Chapter 5
Quick Quiz®

Basic Small Engine Maintenance and Repair Projects

6

INTRODUCTION

In order for OPE to attain its maximum life, a small engine must be properly maintained and repaired when component failures occur. The most basic maintenance and repair projects for small engines can be performed in less than one hour with a few common hand tools. Basic maintenance and repair projects include changing the oil, removing debris, servicing the spark plug and air cleaner, and inspecting and replacing the muffler. Engine troubleshooting and general maintenance can be performed before placing seasonal equipment in long-term storage.

WARNING: When performing any basic maintenance and repair tasks, always wear safety glasses or goggles for protection from flying debris or splashing fuel.

Project: Changing Engine Oil

- **Competency level:** basic
- **Tools needed:** socket wrench set, box wrench or adjustable wrench, screwdriver, hex key, oil filter wrench or pipe wrench (for models with filters), oil drain pan, and funnel
- **Estimated completion time:** 30 min

Fresh oil is a golden or amber color. Gradually, heat, dirt particles, and unburned fuel in the crankcase cause the oil to darken. Dark oil is not only dirty, but it also has lost much of its ability to coat and protect engine components. Small engine OEMs typically recommend changing the oil in a small engine after every 25 hr of operation. For new engines however, oil must be changed after the first 5 hr of operation. New engines require an earlier initial oil change in order to flush out small solid particles that are produced during the engine break-in period.

The type of work the engine performs is equally important. Just as with the oil in a vehicle operated in extremely dirty or dusty conditions or at high speeds, the oil in a lawn mower or other small engine breaks down faster under tough conditions, such as cutting wet and heavy grass, excessive dust, high temperatures, and rough or hilly terrain.

Too much oil can cause the same type of engine damage as not having enough oil. Air bubbles can form in the oil, reducing overall lubrication. The resulting friction and metal-to-metal contact can cause premature part failure. Excess oil can also seep into and burn in the cylinder, producing smoke in the exhaust and leaving excessive carbon deposits in the engine.

To check or replace engine oil, apply the following procedure:

1. Locate the oil fill cap on the crankcase. The location of the oil fill cap varies depending on the make and model of the engine. *See Figure 6-1.* On newer models, look for the oil can symbol or the words "oil" or "fill" stamped on the plug. On lawn and garden tractors, the hood may need to be lifted to locate the cap. Some engines contain either an extended oil fill tube or a standard fill hole with a dipstick for inspection. Others require removing the fill cap to determine if the oil is at the fill line or the top of the fill hole.
2. To prevent dirt and debris from falling into the crankcase, wipe the area around the cap with a clean cloth before removing the cap. If there is no dipstick, dab the oil with the tip of a clean cloth to inspect the oil.
3. If the engine includes a dipstick cap, remove the dipstick and wipe it with a clean cloth. To ensure an accurate reading, reinsert the dipstick completely. Then, remove it again and recheck the oil level. If the dipstick cap is a screw-in type, ensure an accurate reading by screwing it in completely before removing it a second time to recheck the level. The oil mark on the dipstick should be between the full and add lines shown on the dipstick. It should never be above or below these lines. If the oil is dirty, proceed to Step 4.

CHANGING ENGINE OIL . . .

① Locate and remove oil fill cap

② Wipe area around cap with clean cloth
— Oil fill hole
— Clean cloth

③ Remove dipstick and wipe with clean cloth
Full line
Dipstick
Clean cloth

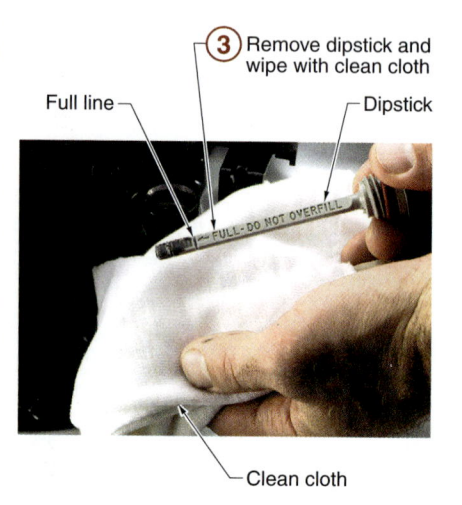

Figure 6-1. Engine oil used in small engines should be changed every 25 hr of operation or once per season.

4. Check the owner's/operator's manual to determine the type and amount of oil required.
5. Run the engine for several minutes. *Note:* Draining the oil while it is warm removes many solid particles that would otherwise settle in the engine.
6. Stop the engine, disconnect the spark plug lead, and secure it away from the spark plug.
7. Locate the oil drain plug. On lawn mowers, the plug is typically below the mower deck and may be obscured by a layer of grass and debris.
8. Wipe the area with a cloth to prevent debris from falling into the crankcase when opening the drain plug.
9. Position an oil pan beneath the mower.
10. Use a socket wrench to rotate the drain plug counterclockwise, allowing the old oil to drain.
11. Replace the drain plug by rotating it clockwise and tightening it with a box wrench, a ¾″ ratchet extension, or adjustable wrench as required.
12. If the engine has an oil filter, replace it by twisting the body counterclockwise, using an oil filter wrench, or pipe wrench.
13. Lightly oil the filter gasket with clean engine oil to promote a tight seal.
14. Install the new oil filter rated for the engine by rotating the filter by hand until the gasket contacts the filter adapter. Tighten the filter an additional ⅛ to ¼ turn.
15. Use a funnel to add the appropriate amount of fresh oil in the crankcase, run the engine at idle speed, and check for leaks.

After an oil change, used oil and soiled rags should be disposed of in accordance with local environmental statutes. In many areas, used oil can be left along a curbside with other recyclables for pick up, provided it is sealed in a recyclable container. *See Figure 6-2.* Rules and regulations should be verified with local municipalities.

Changing Engine Oil

⑩ Remove drain plug and drain oil

⑨ Position oil pan beneath mower

...CHANGING ENGINE OIL

④ Determine oil type
⑤ Run engine for several minutes
⑥ Disconnect spark lead
⑦ Locate oil drain plug
⑧ Wipe oil drain plug area with clean cloth
⑪ Replace drain plug
⑫ Remove old oil filter
⑬ Lightly oil the filter gasket with fresh engine oil
⑭ Install the new oil filter
⑮ Add fresh oil in crankcase

Figure 6-1. (Continued)

DISPOSING OF USED OIL

Recyclable container — Used oil

Figure 6-2. Used oil and soiled rags should be disposed of in accordance to local environmental statutes.

Project: Removing Debris From Engines . . .

- **Competency level:** basic
- **Tools needed:** soft-bristle brush, putty knife, needle-nose pliers, screwdriver, socket set, Torx®-head driver or socket
- **Estimated completion time:** 45 min to 60 min

A temporary loss of power or even permanent engine damage can result once grass or other debris accumulates between the engine parts of a small engine. Debris under the blower housing or in the cooling fins on the cylinder head can cause an engine to run too hot. Prolonged overheating may cause engine damage. Debris can also cause governor linkages to bind or prevent air from reaching the governor blade on a pneumatic governor, resulting in difficulty controlling engine speed.

The blower housing and muffler area should be inspected for debris each time the engine is operated. A good indication that debris has accumulated underneath the blower housing is if the screen over the blower housing is clogged. The blower housing should be removed for a more thorough inspection and cleaning at the end of each season of use and more often if equipment is operated in tall or wet grass. Debris can consist of dry material such as grass, dirt, and chaff.

To remove debris from an engine, apply the following procedure:

1. Disconnect the spark plug lead and secure it safely away from the spark plug.
2. Remove the plastic blower housing. *Note:* If the housing is metal, a set of screws or bolts must be removed. On some models, removing the screws requires a Torx®-head driver or socket.
3. Clean the cooling fins using a soft-bristle brush.
4. Clean the flywheel fins using a soft-bristle brush. Scrape the dirt away gently using a putty knife or brush, being careful not to damage the housing or flywheel. To loosen stubborn grit, apply a light solvent, such as Briggs & Stratton All Purpose Cleaner, to the brush.
5. Clean the flywheel rotating screen using a soft-bristle brush. Remove all debris by hand or with a putty knife and brush.
6. Remove any debris from governor linkages, including the pneumatic governor vane (if equipped), and verify that the linkages move freely. Use a light solvent to loosen remaining dirt and debris.
7. Check for debris around the brake assembly. Verify that the brake cable and linkage move freely.
8. Reattach the blower housing.
9. Reconnect the spark plug wire. *See Figure 6-3.*

Checking Stop Switches

If a lawn mower stops unexpectedly while mowing around trees or bushes, the stop switch wire may have been accidentally disconnected. A stop switch wire is a short wire extending from the brake assembly to the ignition armature. A disconnected stop switch wire may ground the ignition, preventing the spark plug from firing. Under ordinary conditions, the stop switch is designed to stop the engine any time the brake bail is released on the equipment handle. The flexible metal tab on the stop switch should be pressed and the stop switch wire reattached. It is important not to break the wire as it is rotated back into position.

Chapter 6—Basic Small Engine Maintenance and Repair Projects 85

REMOVING DEBRIS FROM ENGINES

- Soft-bristle brush
- ① Disconnect spark plug lead from engine
- ② Remove blower housing
- ③ Clean debris from cooling fins
- ④ Clean debris from flywheel
- ⑤ Clean debris from flywheel screen
- ⑥ Remove debris from governor linkage
- ⑦ Clean debris from brake assembly
- ⑧ Reattach blower housing
- ⑨ Reconnect spark plug wire

Figure 6-3. Several different steps are required to properly remove debris from a small engine.

SMALL ENGINE AND EQUIPMENT MAINTENANCE

...Project: Removing Debris From Engines

To degrease equipment, apply the following procedure:

1. With the equipment in a well-ventilated area and the engine OFF, liberally spray a degreasing agent, such as Briggs & Stratton Heavy-Duty Degreaser, on greasy and dirt-encrusted surfaces.

2. Wait 15 min to allow the grease and dirt to break down. Wipe off the residue with a clean cloth.

3. Use water from a pressurized hose to rinse equipment surfaces. Allow the equipment to dry completely before storing. *See Figure 6-4.*

DEGREASING OPE

① Spray degreasing agent directly on engine

② Wipe off residue with clean cloth

③ Rinse surfaces with water from garden hose

Figure 6-4. Accumulated grease can be removed from OPE by using a degreasing agent and a pressurized water hose.

Project: Servicing Ignition Systems . . .

- **Competency level:** basic
- **Tools needed:** spark tester, spark plug socket (included with most socket sets), socket wrench, wire brush, spray-on plug/point cleaner, rigid putty knife, and spark plug gauge
- **Estimated completion time:** 15 min

Inconsistent firing (spark miss) can result in sluggish engine operation and poor acceleration. The ignition system must be operating properly and the electrodes of a spark plug must be clean and sharp to produce the powerful spark required for ignition. The more worn or dirty a spark plug, the more voltage is required to produce an adequate spark. If the recoil must be pulled repeatedly to start an engine, a worn or damaged spark plug, or faulty ignition component may be the cause. Damaged/worn spark plugs can also cause excessive fuel consumption. In addition, they can cause carbon to be deposited on the cylinder head.

A spark plug is one of the easiest and most inexpensive engine components to service or replace. Although spark plugs can be cleaned, it is recommended to simply replace a dirty, damaged, or worn spark plug.

Replacing Spark Plugs

Spark Plug Location

To service or replace spark plugs, apply the following procedure:

1. Disconnect the spark plug lead, and clean the area around the spark plug to avoid getting debris in the combustion chamber when removing the plug.
2. Remove the spark plug using a spark plug socket wrench.
3. Inspect the spark plug electrodes for hardened deposits. Also, inspect for cracked porcelain or electrodes that have been burned away. Replace the spark plug if any of these conditions exist.
4. Use the spark plug gauge to measure the gap between the two electrodes (one side and one center) at the tip of the spark plug. *Note:* Many small engines require a 0.030″ gap. Check the specifications of the engine model for the gap requirements.
5. Use the gap adjuster tool on the spark plug gauge to bend the side electrode and adjust the size of the gap. Slide the proper gap gauge wire to check gap clearance. When the gap is correct, the gauge will drag slightly between the electrodes as it is pulled through the gap. *See Figure 6-5.*
6. Reinstall the spark plug (being careful not to overtighten), and reconnect the spark plug lead.

SPARK PLUG GAUGES

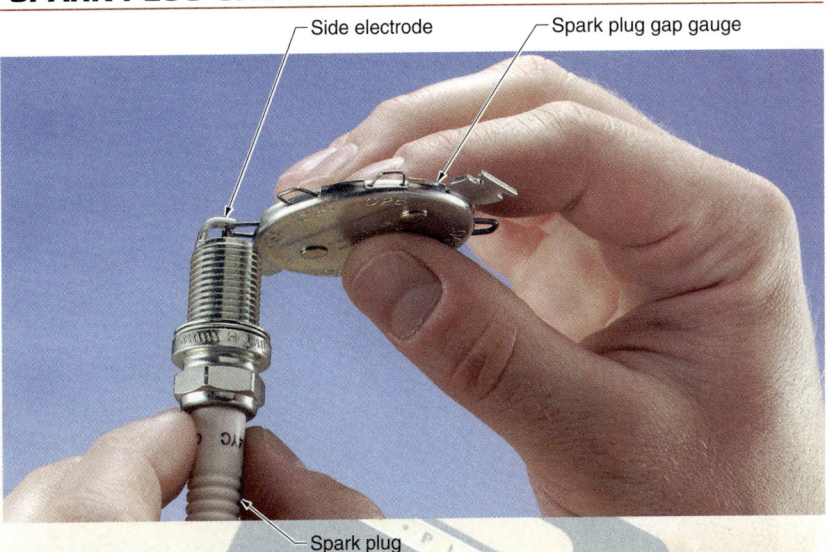

Figure 6-5. A spark plug gauge is used to adjust and measure the gap between two electrodes at the tip of a spark plug.

... Project: Servicing Ignition Systems ...

If inconsistent firing is still present after replacing the spark plug, the rest of the ignition system must be tested. A *spark tester* is a test tool used to test the ignition system of a small engine. *See Figure 6-6.* A spark plug that is fouled or improperly gapped may not allow sparks to jump the gap between electrodes to produce a strong spark. The spark plug will fire erratically. Spark miss can also cause the engine to emit black smoke or produce a popping sound as unburned fuel exits the combustion chamber and ignites inside the muffler.

SPARK TESTERS

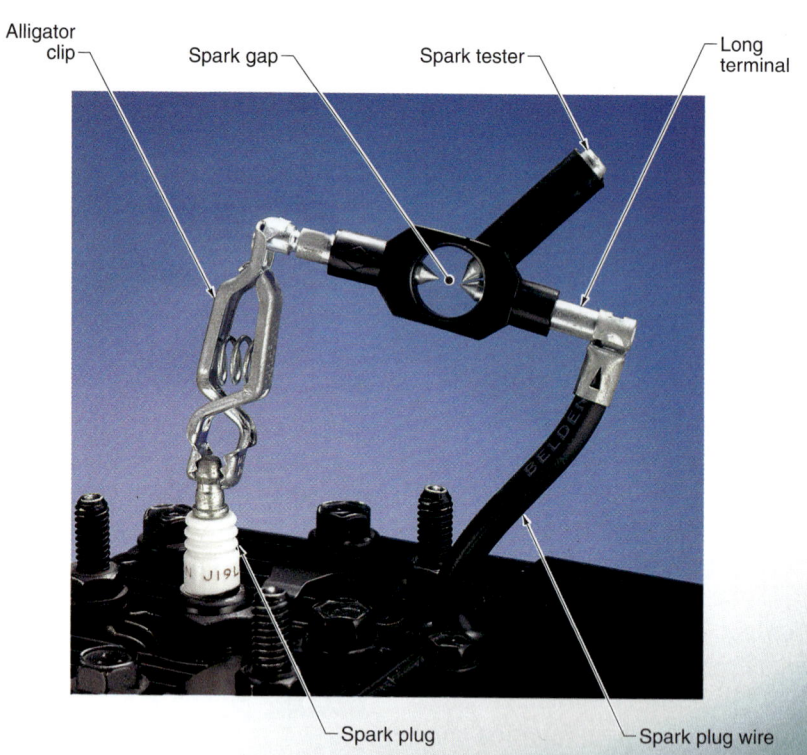

Figure 6-6. A spark tester is used to test the ignition system of a small engine.

To test an ignition system with a spark tester, apply the following procedure:

1. Connect the spark plug lead to the long terminal of the tester. Ground the tester to the engine with the alligator clip of the tester. *See Figure 6-7.*
2. Use the recoil rope or electric starter to crank the engine. Observe if a spark is visible in the spark tester window. If a spark jumps the gap in the tester window properly, the ignition is functioning. The absence of a visible spark indicates a problem in the ignition system, brake, or safety interlock system.

If the ignition system is functioning properly, check the spark plug for spark miss.

To test for spark miss, apply the following procedure:

1. With the spark plug screwed into the cylinder head, attach the spark plug lead to the long terminal of the spark tester.
2. Attach the alligator clip of the spark tester to the spark plug terminal. *See Figure 6-8.*
3. Crank the engine and observe if a spark jumps the gap in the tester window. If an inconsistent spark is present, this can cause the air/fuel mixture to exhaust into the muffler and ignite. If spark miss occurs, the spark plug must be replaced.

TESTING IGNITION SYSTEM WITH SPARK TESTERS

Figure 6-7. A visible spark is present in the window of the tester when the ignition system is functioning properly.

TESTING FOR SPARK MISS

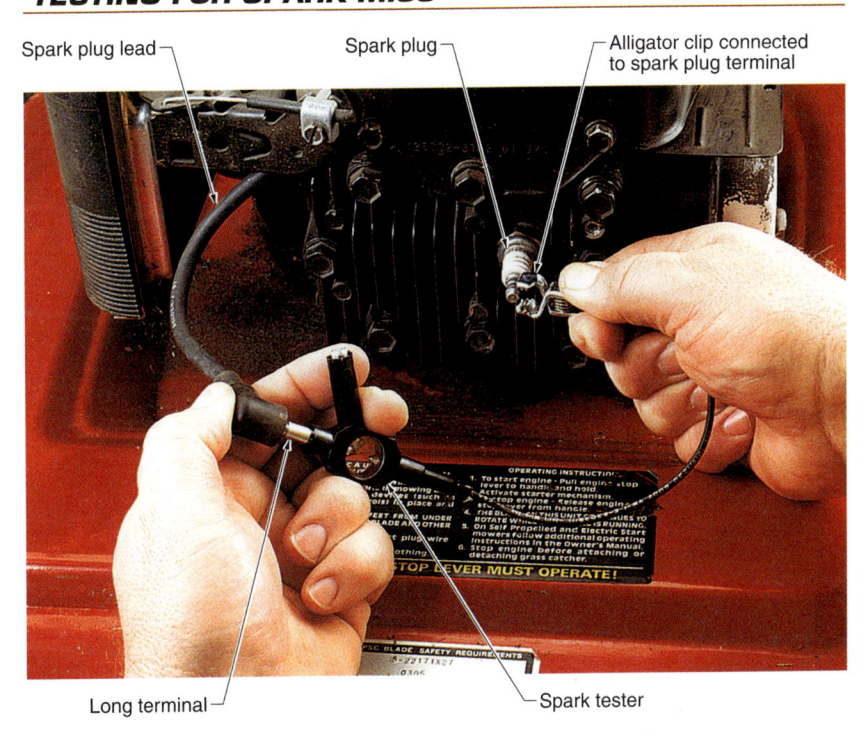

Figure 6-8. A spark plug is tested for spark miss after the ignition system proves to function properly.

Project: Servicing Air Cleaners...

- **Competency level:** basic
- **Tools needed:** screwdriver
- **Estimated completion time:** 15 min

A properly maintained air cleaner is the first layer of defense an engine has against the destructive effects of dirt. An air cleaner in good condition prevents airborne dirt particles from entering through the carburetor. If the air cleaner is not maintained, dirt and dust will gradually enter the engine. Dirt particles in an engine can lead to a sharp drop in engine power or the premature wear of critical engine components. Many types of air cleaners are used in small engines. Most contain an oil foam or pleated paper element.

Air cleaners can be single- or dual-element air cleaners. Single-element air cleaners should be serviced every 25 hr or once a season. The precleaner of a dual-element system should be serviced or replaced every 25 hr. The cartridge should be replaced every 100 hr. The owner's/operator's manual should be used to identify the appropriate air cleaner for the engine. Dual-element air cleaners are available in a variety of designs. *See Figure 6-9.*

Dual-element air cleaners contain a pleated paper element with a foam precleaner, thereby offering two layers of protection. When dirt can no longer be removed from the pleats by tapping the element on a hard, dry surface, the paper element should be discarded. Although dirt can be removed from some oil-foam precleaner elements with hot water and liquid dish detergent that contains a grease-cutting agent, it is recommended to replace the elements when dirty.

DUAL-ELEMENT AIR-CLEANER DESIGNS

Figure 6-9. Dual-element air cleaners are available in a variety of designs.

... Project: Servicing Air Cleaners

To service oil-foam air cleaners, apply the following procedure:

1. Use a screwdriver to loosen the attachment screws (or wing nuts) that hold the cover of the foam air cleaner in place. **See Figure 6-10.** Disconnect the spark plug lead and secure it safely away from the spark plug.
2. Disassemble the air cleaner.
3. Inspect the foam air-cleaner element. Replace it if it is torn or shows signs of wear.
4. Saturate the new oil-foam air-cleaner element with engine oil and squeeze it to spread the oil throughout. **See Figure 6-11.**
5. Inspect the rubber sealing surface between the air cleaner and carburetor. Replace element if it is torn or shows signs of wear.
6. Reassemble and reinstall the air cleaner element. Reconnect the spark plug lead to the spark plug.

REMOVING AIR-CLEANER ASSEMBLIES

Figure 6-10. An attachment screw must be loosened to remove the air cleaner assembly cover on most small engines.

FOAM AIR-CLEANER ELEMENTS

Figure 6-11. Foam air-cleaner elements must be saturated with engine oil prior to installation.

DUAL-ELEMENT AIR CLEANERS

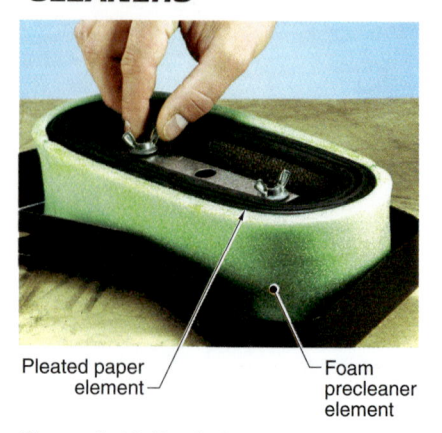

Figure 6-12. Dual-element air cleaners contain foam precleaner and a pleated paper element.

To service a dual-element air cleaner element, apply the following procedure:

1. With the air cleaner cover removed, separate the foam precleaner (if equipped) from the paper element and tap it gently on a flat surface to remove any loose dirt. Inspect the element and replace it if it has been used for 100 hr or is heavily soiled, wet, or crushed. **See Figure 6-12.** Disconnect the spark plug lead and secure it safely away from the spark plug.
2. Inspect the foam precleaner. Note the mesh backing designed to act as a barrier between the precleaner and the pleated paper element. Replace it every 25 hr or when soiled.

3. Look for installation instructions on the foam precleaner. *See Figure 6-13.* If directed, lubricate the foam precleaner with oil. *Note:* Some foam precleaners are not required to be oiled.
4. Clean the air cleaner element housing using a clean, dry cloth. Do not clean with solvents or compressed air. Replace the paper element if dirty. *See Figure 6-14.*
5. Reassemble the air cleaner components. If instructed to oil the precleaner, insert the mesh toward the paper element so that the paper is never directly exposed to the oil.
6. Reinstall the air cleaner element, verifying that any tabs on the air cleaner housing are in their slots on the carburetor backing plate. Gaps around the element will permit unfiltered air and damaging dirt particles to enter the engine. Reconnect the spark plug lead to the spark plug.

TECH FACT

If an engine has a paper filter element, it should be removed temporarily any time the engine must be tipped on its side. This will eliminate any chance that oil from the precleaner or the crankcase will spill onto the paper and destroy it. To prevent debris from entering the carburetor, the carburetor opening should be temporarily covered with a sheet of plastic.

PRECLEANER INSTALLATION INSTRUCTIONS

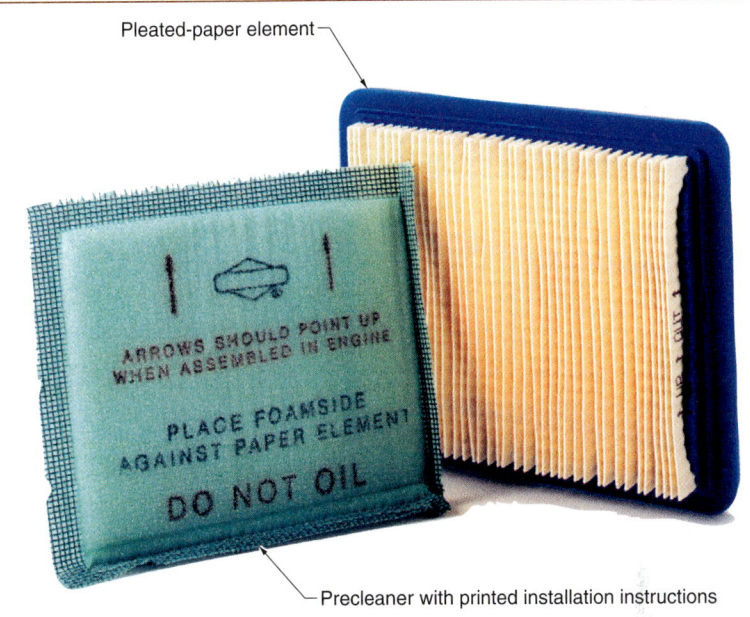

Figure 6-13. Many precleaners have installation instructions printed directly on the surface.

CLEANING AIR CLEANER CARTRIDGE HOUSINGS

Figure 6-14. Air-cleaner element housings should be cleaned with a clean, dry cloth instead of solvents or compressed air.

Project: Inspecting and Replacing Mufflers

- **Competency level:** basic
- **Tools needed:** hammer, rubber mallet, pin punch, slip-joint pliers, socket wrench, and penetrating oil such as WD-40
- **Estimated completion time:** 30 min

One of the main causes of small engine noise is hot gas that is forced out of the cylinder during each exhaust stroke. A muffler is used to reduce engine noise. It also functions as a spark arrestor by preventing exhaust sparks from exiting and reducing the possibility of igniting dry grass, leaves, or other debris. However, after a season or two of use, exhaust gases leave a layer of hard soot in the muffler. This layer reduces the inside wall size of the muffler and creates additional resistance to gases exiting the cylinder, causing the muffler to operate improperly. A muffler must be replaced if there is an accumulation of hard soot or excessive noise. A muffler must also be replaced if cracks or holes are present.

Engine noise is caused by the sound waves produced when exhaust gases rush through the small opening of the exhaust valve. Engine noise is reduced when the exhaust gases are routed in the muffler through a series of perforated baffles and plates that disperse sound waves. A damaged muffler is easily noticed by the dramatic increase in engine noise.

To inspect and change a muffler, apply the following procedure:

1. Locate the muffler, which is usually near the cylinder head. Disconnect the spark plug lead and secure it safely away from the spark plug.
   CAUTION: Allow a muffler to cool completely before handling. The surface of a muffler can remain hot enough to burn skin shortly after the engine is shut off. In addition, a rusty, cracked muffler can cut skin. Therefore, wear leather work gloves and use slip-joint pliers to remove a muffler.
2. Inspect the exterior of the muffler for soot, rust, dents, holes, or cracks. If any of these conditions are visible, proceed to Step 3.
3. Remove the muffler. *Note:* A muffler may be attached directly to an engine with mounting bolts or screwed into the exhaust port. With some mufflers, an extended pipe threads into the engine.
   a. If the muffler is attached with mounting bolts and has locking tabs around the bolts, use slip-joint pliers to bend the locking tabs back far enough to fit a socket wrench over the bolt heads. Remove the bolts with a socket wrench and detach the muffler. *See Figure 6-15.*
   b. If the muffler is threaded and screwed directly into an exhaust port, apply penetrating oil to the screw threads. *See Figure 6-16.* Allow the oil to soak for several minutes. Then, grasp the screws with slip-joint pliers and rotate counterclockwise.

REMOVING BOLTED MUFFLERS

Figure 6-15. A muffler that is attached to an engine with bolts can be removed by loosening the bolts with a socket wrench.

REMOVING THREADED MUFFLERS

Figure 6-16. A muffler that is screwed directly into an exhaust port should have penetrating oil applied to the screw threads prior to removal.

Chapter 6—Basic Small Engine Maintenance and Repair Projects

c. If the muffler is attached with a threaded lock ring, loosen the lock ring by tapping it with a hammer and pin punch. Then, grasp the muffler with slip-joint pliers and avoiding damaging the threads, rotate it counterclockwise. **See Figure 6-17.**

4. Dislodge hard soot accumulation in the muffler by tapping the muffler body with a rubber mallet or on a hard surface. **See Figure 6-18.** If the muffler is in good condition, proceed to Step 5. If the muffler is damaged or hard soot cannot be removed, proceed to Step 6.

5. Reattach the muffler. *Note:* Before reattaching a muffler, brush dirt and debris off the threaded area. If left on the muffler, dried grass clippings and other debris can ignite on the hot surface of the muffler. Do not overtighten bolts or screws when reattaching the muffler.

6. Replace the muffler with OEM parts. Reconnect the spark plug lead to the spark plug.

Removing Mufflers

REMOVING MUFFLERS WITH THREADED LOCK RINGS

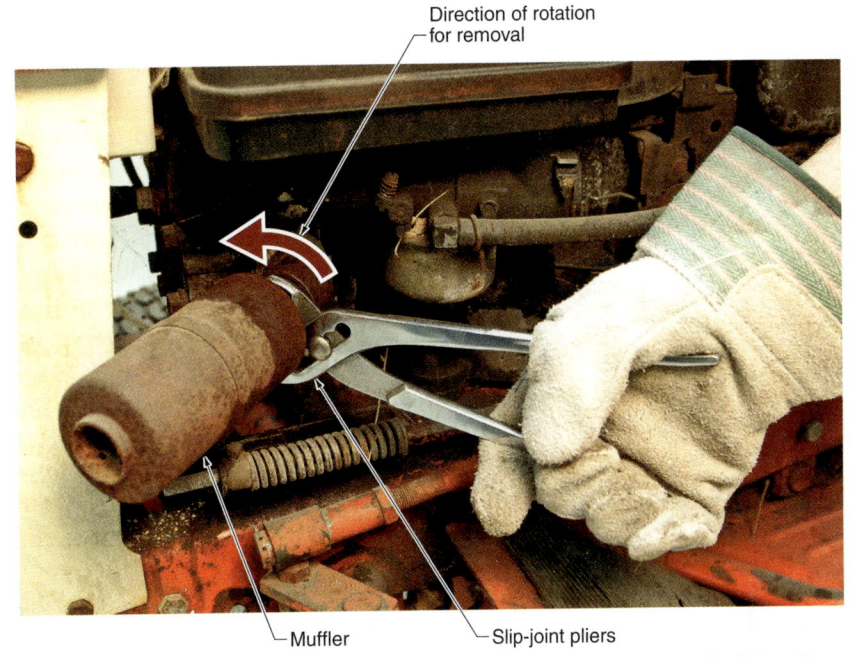

Figure 6-17. A threaded muffler can be removed by grasping with slip-joint pliers and rotating counterclockwise.

HARD SOOT REMOVAL

Figure 6-18. Hard soot accumulation in a muffler can be removed by tapping the muffler body with a rubber mallet or on a hard surface.

SMALL ENGINE AND EQUIPMENT MAINTENANCE

Project: Replacing Mower-Deck Drive Belts

- **Competency level:** basic
- **Tools needed:** socket or open-end wrenches
- **Estimated completion time:** 30 min

The most common drive system for mower implements is the mower-deck drive belt. These belts require replacement at regular intervals. The most common indications of belt problems are slippage while damp or wet and a squealing noise upon engagement. The proper operation of the mower-deck drive belt is important for the efficient operation of any lawn cutting equipment.

Sketching Belt Routes

To inspect and replace a mower-deck drive belt, apply the following procedure:

1. Place the equipment on a hard, level surface such as a concrete driveway or garage floor.
2. Remove the spark plug wire and secure it away from the spark plug.
3. If the unit has a manual mower-deck engagement lever, move it to the disengaged position.
4. Move the height adjustment lever to the lowest possible cutting height. If so equipped, release the quick-disconnect mechanism, slide the belt from the main engine to the deck drive off the engine drive pulley, and slide the mower deck out from the side and away from the equipment. ***See Figure 6-19.***
5. Examine the mower-deck drive belt or belts for discoloration, cracks, or tears in the belt body.
6. Examine the belt routing and sketch the route on paper for reference.
7. Remove any belt covers that may block the view of the entire belt. *Note:* If the mower deck is equipped with partially or fully enclosed belt guards, loosen them with a socket or open-end wrench so that the belt can be easily removed from the pulleys. If the mower deck is equipped with stationary belt guide bolts, loosen them also. Retract any spring-loaded pulleys that may exert pressure on the belts.
8. Remove the old belt.
9. Install the new belt using the same routing as sketched previously.
10. Reset the belt guides so that a minimum of ¼″ of clearance is maintained between the guide and flat surface of the properly tensioned belt.
11. Use the socket or open-end wrench to tighten the belt guides.
12. Replace the belt covers on the mower deck.
13. Reinstall the mower deck under the equipment by reversing the assembly. *Note:* If the unit does not have a quick-disconnect mower deck, manually position the lift arm to the desired cutting height. If the unit has a manual engagement lever, move the lever to the engaged position and inspect the belt tension and clearance at the belt guides. If the unit has an electric clutch engagement system, the mower-deck drive belt will be under constant tension when installed. Slip the main engine deck drive belt back in place. Visually inspect for clearance between the belt guides and flat side of the mower-deck drive belt.
14. Reconnect the spark plug wire to the spark plug and return the equipment to service.

QUICK-DISCONNECT MOWER DECKS

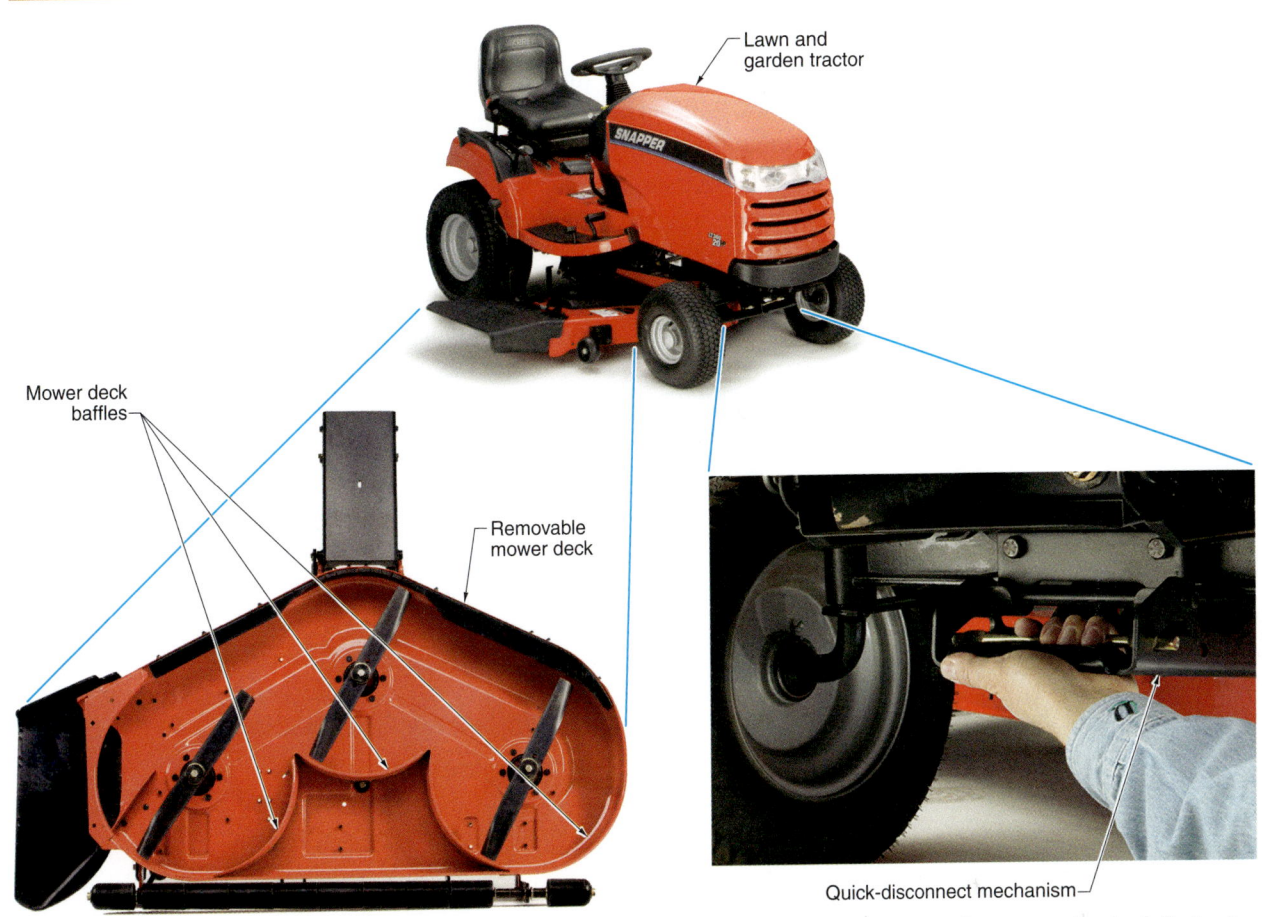

Figure 6-19. Some lawn and garden tractors are equipped with mower decks that can be easily removed or installed using a quick-disconnect mechanism.

Replacing
Mower-Deck
Drive Belts

Removing Lawn
Mower Blades
with a Lawn
Mower Blade
Locking Tool

Project: Replacing Snow Thrower Shear Pins

- **Competency level:** basic
- **Tools needed:** socket or open-end wrenches, ¼″ pin punch, and small hammer
- **Estimated completion time:** 15 min

Every two-stage snow thrower has shear pins in the auger system. A shear pin is a solid cylinder of metal alloy that fractures or shears at a predetermined location and applied torque value. Shear pins are designed to break when a certain amount of force is applied to a rotating auger shaft. However, they are simple to replace.

Typically, when the auger strikes a solid object, the shear pins fracture, greatly reducing the chance of damage to the auger, auger shaft, and auger drive transmission. Most shear pins are designed similarly to standard bolts, however, with serrations cut in the body of the pin where the fracture is to occur. *See Figure 6-20.*

To inspect and replace snow thrower shear pins, apply the following procedure:

1. Place the snow thrower on a hard, level surface such as a concrete driveway or garage floor.
2. Remove the spark plug wire and secure it safely away from the spark plug.
3. Visually inspect the auger shafts by rotating each auger slowly. *Note:* If only one shear pin has been fractured, at least one auger will rotate freely. While slowly rotating the freely rotating auger, visually inspect for a hole in the shaft where the shear pin was located. Align the auger with the hole in the auger shaft.
4. Using a ¼″ pin punch and small hammer, tap the fractured shear pin out of the auger shaft, leaving the pin punch in place.
5. Gently remove the pin punch and insert a new shear pin into the hole in the auger, through the auger shaft, and out the opposite side.
6. Use a socket or open-end wrench to install and tighten the nut on the opposite side of the shear pin head.
7. Reconnect the spark plug wire to the spark plug and return the equipment to service.

   CAUTION: Remove and properly dispose of fractured shear pin pieces before starting the snow thrower.

SHEAR PINS

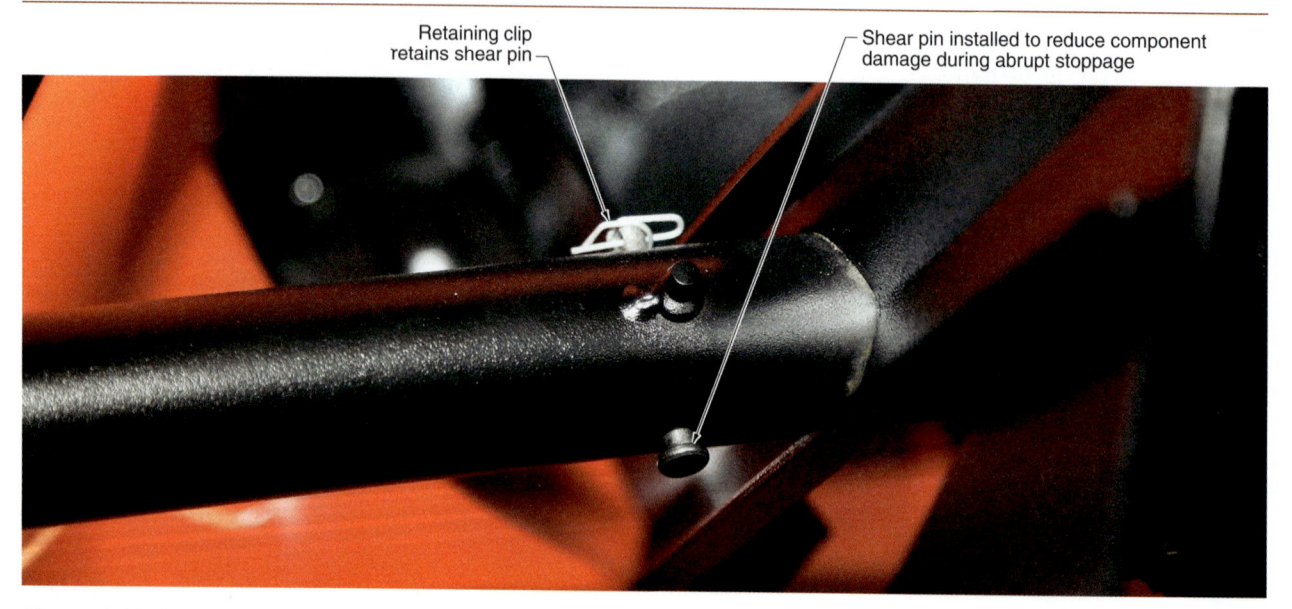

Figure 6-20. Shear pins are designed to reduce component damage by fracturing when the driven member of a rigid system requires more power than a disc or coupling is able to provide.

Project: Replacing Snow Thrower Skid Shoes

- **Competency level:** basic
- **Tools needed:** socket, box, adjustable, or open-end wrenches; minimum size 2″ × 4″ × 30″ wooden board; and one pair of wooden paint mixing sticks
- **Estimated completion time:** 15 min

Two-stage snow throwers typically have skid shoes to help provide smooth movement of the snow thrower along uneven surfaces. Due to the heavy weight of the snow thrower, skid shoes are subject to wear over a period of time and must be replaced. *See Figure 6-21.*

To replace the skid shoes on a snow thrower, apply the following procedure:

1. With the snow thrower on a smooth, level surface, chock the wheels. Then, place a wide wooden board several inches thick, or other object of similar thickness, under the scraper bar. *Note:* The board or other object must be thick enough to suspend the skid shoes off the surface.
2. Using the proper size wrench, remove the retaining bolts from the skid shoes. Then, remove the skid shoes from the snow thrower.
3. With the machine on a smooth, level surface, install the new skid shoes and retaining bolts, but do not completely tighten the retaining bolts.
4. Remove the board or other object from under the scraper bar, and replace with a pair of wooden paint mixing sticks or cardboard of similar thickness.
5. Adjust the skid shoes to make full contact with the ground surface.
6. Use the proper size wrench to tighten the retaining bolts. *Note:* If the snow thrower is to be used on any type of surface other than concrete or asphalt, the skid shoes may need to be adjusted to a height higher than normal to prevent them from collecting debris, such as small stones or dirt, when the snow thrower is in use.

SNOW THROWER SKID SHOES

Figure 6-21. Two-stage snow throwers typically have skid shoes to help provide smooth movement of the snow thrower along uneven surfaces.

Project: Engine Troubleshooting and General Maintenance

- **Competency level:** basic
- **Tools needed:** socket wrench set, needle-nose pliers, and feeler gauge
- **Estimated completion time:** 15 min to 45 min

Engine troubleshooting and general maintenance include testing engine compression, inspecting the crankcase breather, lubricating cables and linkages, and tightening bolts. These tasks are performed as part of seasonal maintenance or any time an engine has operated under a heavy load or dirty or dusty conditions.

Each maintenance task helps to ensure proper engine operation and optimize engine performance. The engine compression is tested to verify the maximum power output of each piston stroke. The crankcase breather is inspected to ensure that crankcase gases are properly vented. The cables and linkages should be lubricated frequently during the season to avoid the minor problems associated with binding controls or a sluggish governor. Tightening bolts is necessary to ensure safe equipment operation and to protect the engine block and equipment body.

When an engine has leakage around the valves or rings, compression of the air-fuel mixture is reduced. When this occurs, engine performance and efficiency can drop dramatically. A spin of the flywheel will determine whether the compression in the engine is sufficient. Many engines contain a crankcase breather to vent gases that accumulate in the crankcase. The breather, if the engine is equipped with one, is usually located over the valve chamber.

To test engine compression, apply the following procedure:

1. Disconnect the spark plug lead and secure it safely away from the spark plug.
2. Remove the blower housing and temporarily secure the operator presence bail on the mower handle.
3. Rotate the flywheel counterclockwise by hand. If compression is adequate, the flywheel should rebound sharply. Weak or nonexistent rebound indicates poor compression. ***See Figure 6-22.***

TESTING ENGINE COMPRESSION

Figure 6-22. Testing engine compression verifies the maximum power output of each piston stroke.

To test a crankcase breather, apply the following procedure:

1. Remove the muffler or other parts as needed to reach the crankcase breather. Disconnect the spark plug lead and secure it safely away from the spark plug.
2. Loosen the breather retaining screws and remove the crankcase breather.
3. Verify proper breather valve operation by using a feeler gauge to check the gap between the fiber valve and the breather body. If a 0.045″ feeler gauge can be inserted, replace the breather. *See Figure 6-23.* Reconnect the spark plug lead to the spark plug.

Note: Avoid using force or pressing on the fiber disc, and never disassemble the breather. If the breather is damaged, replace it. Replace the old gasket that fits between the breather and the engine body whenever the breather is removed.

Tightening Bolts

Engine bolts must be tight at all times. If bolts are loose, parts can be damaged easily during engine operation. Mounting bolts that attach the engine to equipment can also loosen, leading to damage such as a cracked engine block. All accessible nuts and bolts should be inspected during regular maintenance. Some mounting bolts must be grasped with a wrench from above and tightened with a second wrench from beneath the equipment. Other bolts are self-tapping and overtightening of the bolts can damage the threads. An authorized dealer should be consulted to determine the proper torque for each type of bolt on a specific engine. A torque wrench should be used for final tightening.

TESTING CRANKCASE BREATHERS

① Remove muffler and other parts as needed to access crankcase location

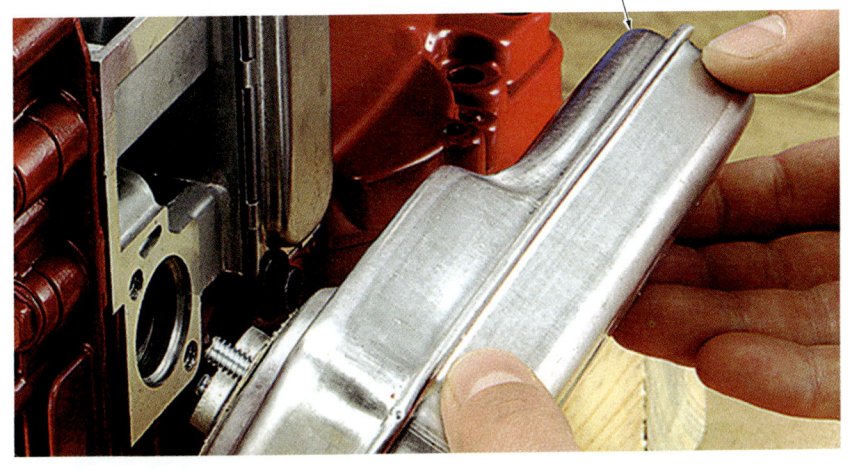

② Remove crankcase breather

③ Use feeler gauge to check gap between fiber valve and feeler body

Figure 6-23. Testing a crankcase breather ensures that crankcase gases are properly vented.

TECH FACT

Control cables and linkages on the flywheel brake and throttle can seize and may even negatively affect engine performance if they cannot move freely. To prevent problems, binding should be reduced to a minimum. Also, cables and linkages should be sprayed occasionally with a solvent or lubricant to keep them free of dirt and debris.

Project: Preparing Equipment for Long-Term Storage...

- **Competency level:** basic
- **Tools needed:** socket or open-end wrenches, putty knife or scraper, grease gun, soft-bristle brush, wire brush, and sandpaper
- **Estimated completion time:** 30 min

Each piece of equipment may have different long-term storage requirements, which are defined in the owner's/operator's manual provided by the OEM. However, there are general guidelines that apply to all OPE. Generally, to properly store OPE such as a lawn mower, the required engine, mowing implement, and battery storage procedures must be followed.

To prepare a rotary lawn mower for long-term storage, apply the following procedure:

1. After running any remaining fuel out of the mower, place the rotary lawn mower on a hard, level surface such as a concrete driveway or garage floor.
2. Remove the spark plug wire and secure it safely away from the spark plug.
3. Tilt the lawn mower up on its rear wheels, examine under the mower deck, and inspect the area for debris and engine oil leakage. If the lawn mower is equipped with a self-propelled flexible belt drive, check the belt also.
4. Clean the underside of the mower deck of all debris using a putty knife or similar scraping tool.
5. Place the lawn mower flat on all four wheels and examine the wheel adjusters for damage. Lightly lubricate any moving parts with petroleum-based oil or spray lubricant.
6. Actuate any cables, speed control mechanisms, and other controls to ensure proper operation. Lightly lubricate as needed with spray lubricant, and then actuate each lubricated component to spread the lubricant evenly onto the moving parts. *See Figure 6-24.*
7. Place the lawn mower in the storage location. Cover the engine with a thin plastic sheet to prevent dust, dirt, and other debris from entering the engine area.

SPRAY LUBRICANTS

Figure 6-24. When preparing a lawn mower engine for long-term storage, the moving components should be lubricated lightly with spray lubricant. Each lubricated component should then be actuated to spread the lubricant evenly into the moving parts.

To prepare a ride-on lawn mower implement for long-term storage, apply the following procedure:

1. Remove the spark plug wire and secure it safely away from the spark plug.
2. If the mower deck is a quick-disconnect mower deck, remove the main drive belt from the engine drive pulley and disconnect the mower deck from the equipment. Slide the mower deck out from the equipment to gain access to moving parts. If the mower deck is standard, lower the mower deck to the lowest cutting position to gain access to moving parts.
3. Remove any belt covers from the top of the mower deck.
4. Clean the outer surface of the mower deck of all debris.
5. Examine the belt drive system of the blade for worn, damaged, or misaligned belts. *See Figure 6-25.*
6. Release the tension idler pulley by moving it away from the tensioning spring to relieve any belt tension. *Note:* On many quick-disconnect mower decks, there is an engine-to-deck flexible belt drive and a flexible-blade belt drive that includes an independent tension idler pulley.
7. Remove the flexible-blade belt drive from at least one of the blade drive pulleys to ensure that the belt is relaxed.
8. For a non-quick-disconnect mower deck, locate the manually actuated tension idler pulley and verify that it is in the release position.
9. Lubricate each blade bearing with a grease gun if the blade bearings are equipped with zerk grease fittings.
10. Lubricate the tension idler pulley with a small amount of spray lubricant.
11. Tilt the mower deck up to examine the underside or cutting area of the deck.
12. Clean the underside of the mower deck of all debris using a putty knife or similar scraping tool.
13. Examine the blade bearings for damage and locate any zerk grease fittings in the bearing housing.
14. Grease each blade bearing with a grease gun.
15. Place the mower deck in the storage location and replace plug wire.

BELT-DRIVE-SYSTEM STORAGE

Figure 6-25. Prior to long-term storage, the belt drive systems of mower deck blades should be examined for worn, scorched, or misaligned belts.

... Project: Preparing Equipment for Long-Term Storage

When practical, the 12 V lead-acid batteries of ride-on OPE should be removed from the equipment and properly stored.

To prepare a battery of OPE for long-term storage, apply the following procedure:

1. Gain access to the battery, which is typically located under the hood or seat of the equipment.
2. Use a soft-bristle brush to clean any deposits or debris from the top of the battery case.
3. Remove the black (NEG) (–) cable and place the cable end away from the battery.
4. Remove the red (POS) (+) cable and place the cable end away from the battery.
5. Identify and loosen any battery retainer device, and then lift the battery from its tray.
6. Clean each battery terminal connector using a wire brush.
7. Clean each battery cable end on the equipment using a wire brush and sandpaper.
8. Place the battery in a dry area away from any spark or open flame for long-term storage.

Chapter 6 Quick Quiz®

Many modern ride-on lawn mowers have drive systems that require some level of service before being placed in long-term storage. If driven by a hydrostatic drive system, the lawn mower can be safely stored without any preparation. If the equipment is driven by a flexible belt drive system or a disc drive system, some storage preparation is required.

To prepare a lawn mower driven by a flexible belt drive system or a disc drive system for long-term storage, apply the following procedure:

1. Remove any spark plug wires and secure safely away from the spark plugs.
2. Locate the tensioner pulley and retract it away from the belt. Slip the belt out from under the pulley to relieve tension. Usually, there is an idler pulley connected with linkage to the clutch pedal. **See Figure 6-26.**
3. If the lawn mower uses a disc drive system, tilt up the lawn mower to gain access to the disc. Place the speed control lever in the neutral position. Verify that the driven disc is no longer in contact with the drive disc.
4. Lubricate any drive speed linkages with spray lubricant.
5. Place the equipment in the neutral position, and then place in the storage location.

IDLER PULLEY AND CLUTCH PEDAL

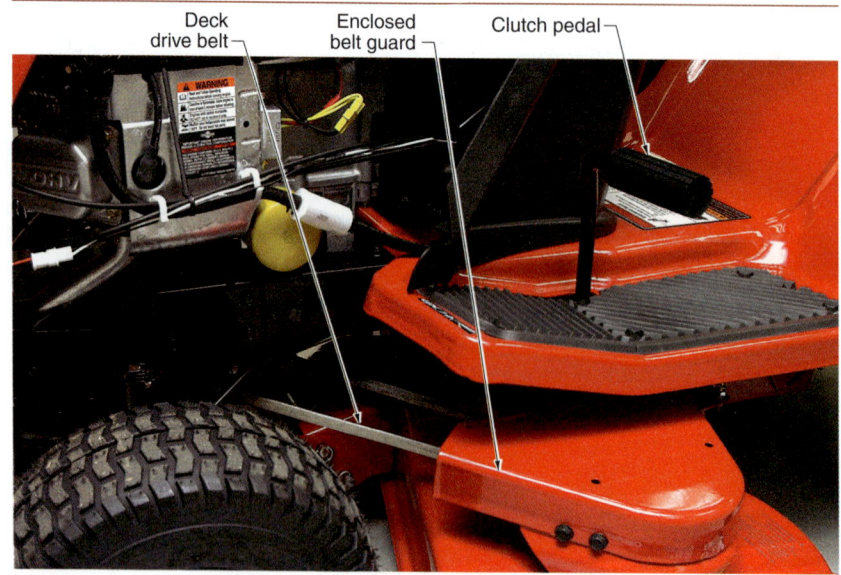

Figure 6-26. There is often an idler pulley connected with linkage to the clutch pedal of ride-on lawn mowers.

Intermediate Small Engine Maintenance and Repair Projects

7

INTRODUCTION

In order for OPE to attain its maximum life, a small engine must be properly maintained and repaired when damage occurs. Intermediate small engine maintenance and repair projects typically require more time and effort than basic projects. Many intermediate maintenance and repair projects for small engines can be performed in less than 90 min with a few common and specialty small engine hand tools. Intermediate maintenance and repair projects include carburetor overhaul, flywheel maintenance, flywheel brake service, fuel system service, governor system service, and rewind starter service. Most intermediate projects require the adjusting and testing of components once they have been repaired.

WARNING: When performing any intermediate maintenance and repair project, always wear safety glasses or goggles for protection from flying debris or splashing fuel.

SMALL ENGINE AND EQUIPMENT MAINTENANCE

Project: Overhauling Carburetors...

- **Competency level:** intermediate
- **Tools needed:** fuel line clamp, box wrench set, 5/32" pin punch, compressed carburetor cleaner or parts cleaning solvent, compressed air, carburetor repair kit, small hammer, screwdrivers (flat head or jet), torque wrench, paper, and pencil or pen
- **Estimated completion time:** 90 min

Most carburetor problems are caused by dirt particles, varnish, and other deposits that block the narrow fuel and air passages. Old gaskets and O-rings are also common sources of problems because they eventually shrink, causing fuel and air leaks that lead to poor engine performance. When overhauling a carburetor, a carburetor repair kit, which includes replacement gaskets and other necessary parts, or a replacement carburetor is required. Also required are carburetor cleaner, a clean work surface, and a source of compressed air for blowing out loosened debris and solvent.

Carburetor cleaner is a powerful solvent that can harm carburetor parts, especially plastic parts if exposed to the cleaner for long periods of time. A carburetor should be soaked with carburetor cleaner for no more than 15 min. Flexible rubberized parts, such as seals, O-rings, and pump diaphragms, should never be exposed to carburetor cleaner. Only compressed air and carburetor cleaner should be used to clear clogged passages or tiny meter holes in the carburetor. Carburetor parts can be easily viewed after the carburetor is removed from the engine. *See Figure 7-1.*

Overhauling a carburetor involves removing the carburetor, disassembling the float-type carburetor, cleaning and inspecting the carburetor and idle mixture screw, and reattaching the carburetor and air-cleaner assembly. **CAUTION:** Carburetor holes and passages should never be reamed or drilled because they are precisely sized and can be permanently damaged if any solid material, such as a wire or drill bit, is inserted in them.

CARBURETOR PARTS

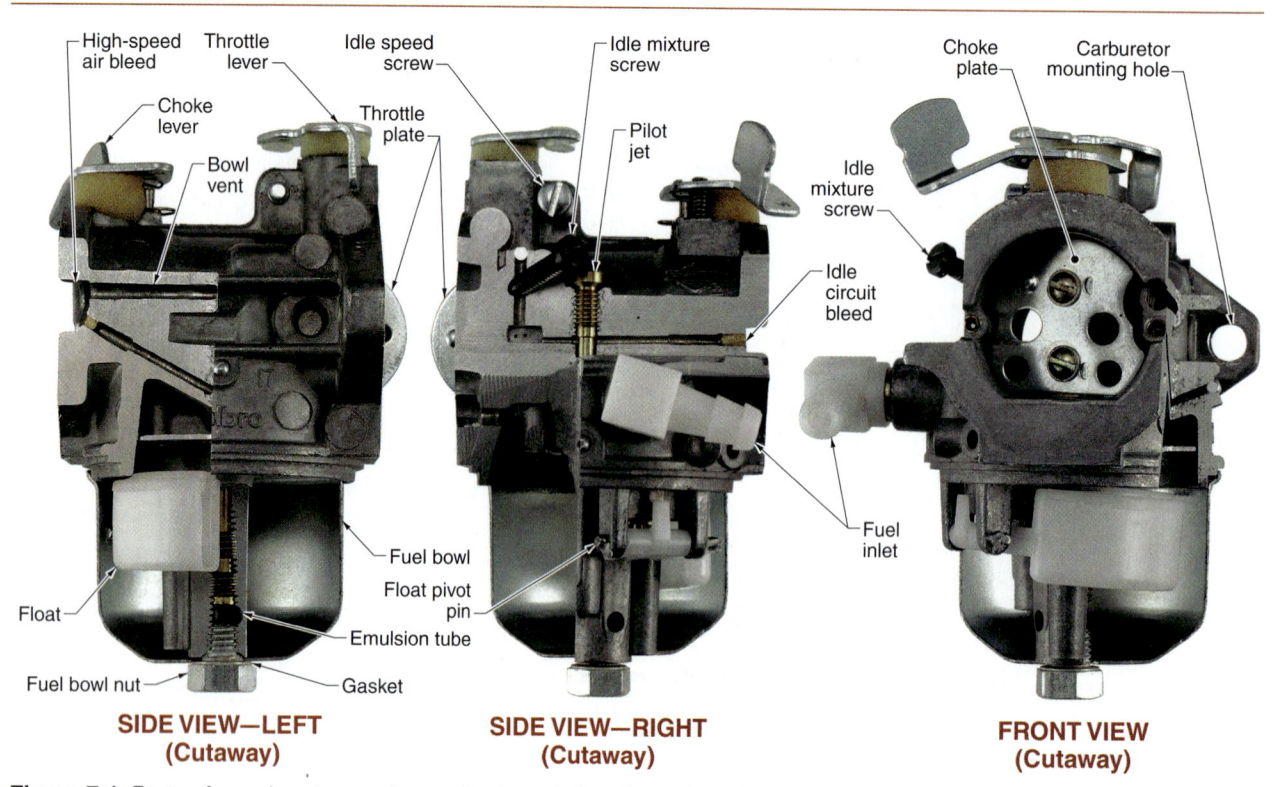

Figure 7-1. Parts of a carburetor can be easily viewed after the carburetor has been removed from the engine.

To remove a carburetor, apply the following procedure:

1. Disconnect the spark plug lead and secure it away from the spark plug. Remove the air cleaner assembly.
2. Shut off the fuel valve at the base of the fuel tank. *Note:* If the engine does not contain a fuel valve, use a fuel line clamp to prevent fuel from draining out of the tank while the carburetor is disconnected from the engine. If the engine has a solenoid, proceed to Step 3. Otherwise, proceed to Step 4.
3. Disconnect the solenoid by removing the wire connector from the solenoid receptacle (some carburetors contain a solenoid at the base of the fuel bowl to control after-fire).
4. With the carburetor still connected to the governor, use a box wrench to unfasten the carburetor mounting bolts. Disconnect the carburetor from the intake manifold elbow by removing the nuts and sliding the carburetor off the studs.
5. Sketch the governor spring and linkage positions on a sheet of paper before disconnecting them (simplifies reattachment).
6. Disconnect the governor springs and linkage, and remove the carburetor, being careful not to bend or stretch the links, springs, or control levers. **See Figure 7-2.**

REMOVING CARBURETOR

① Disconnect spark plug and remove air-cleaner assembly

② Shut off fuel valve or clamp fuel line

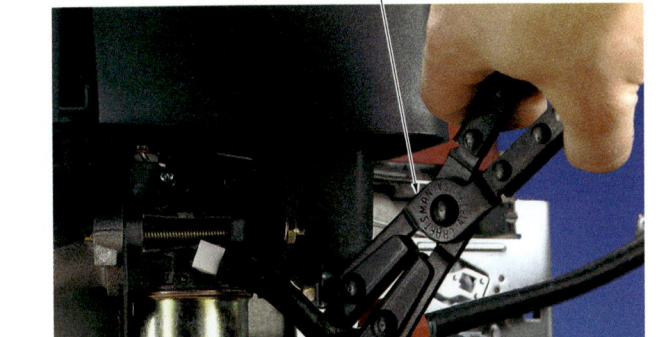

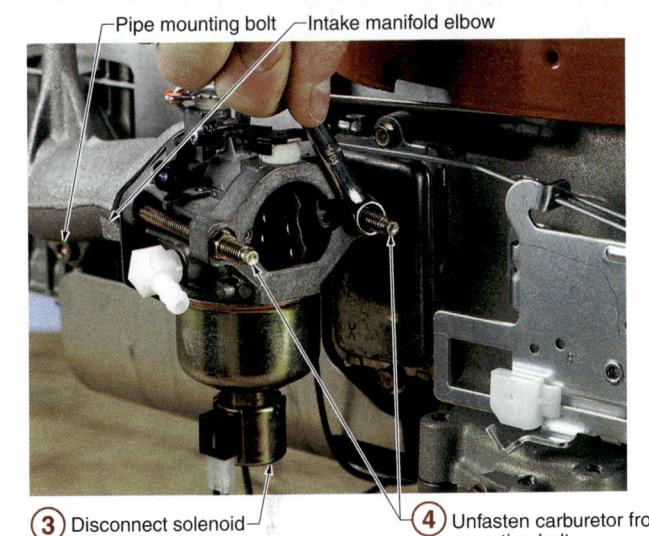

Pipe mounting bolt — Intake manifold elbow

③ Disconnect solenoid

④ Unfasten carburetor from mounting bolts

⑤ Sketch governor springs and linkage; remove carburetor

⑥ Disconnect governor springs and linkage; remove carburetor

Figure 7-2. Removing a carburetor requires removing other components, including the air-cleaner assembly, solenoid, and governor linkage.

...Project: Overhauling Carburetors...

To disassemble a float-type carburetor, apply the following procedure:

1. Remove the fuel bowl from the carburetor body. The fuel bowl may be attached with either a bolt or high-speed mixture screw. *Note:* A fuel bowl may contain a small amount of fuel. A clean bowl should be used to catch dripping fuel and to store small parts. During disassembly, the bowl should be inspected for dirt and debris to determine the general condition of the carburetor.
2. Push the hinge pin out of the carburetor body with a 5/32″ pin punch. Tap only the pin to avoid damaging the carburetor body.
3. Remove the float assembly, inlet needle valve, and fuel bowl gasket.
4. Remove the idle mixture screw along with the spring.
5. Unscrew the main jet from the side of the carburetor pedestal (if equipped), and unscrew the emulsion tube. *Note:* A carburetor (jet) screwdriver is designed to fit into the slot in the head of the emulsion tube so that the threads inside the pedestal or the tube itself will not be damaged while loosened.
6. Remove the emulsion tube. **See Figure 7-3.**

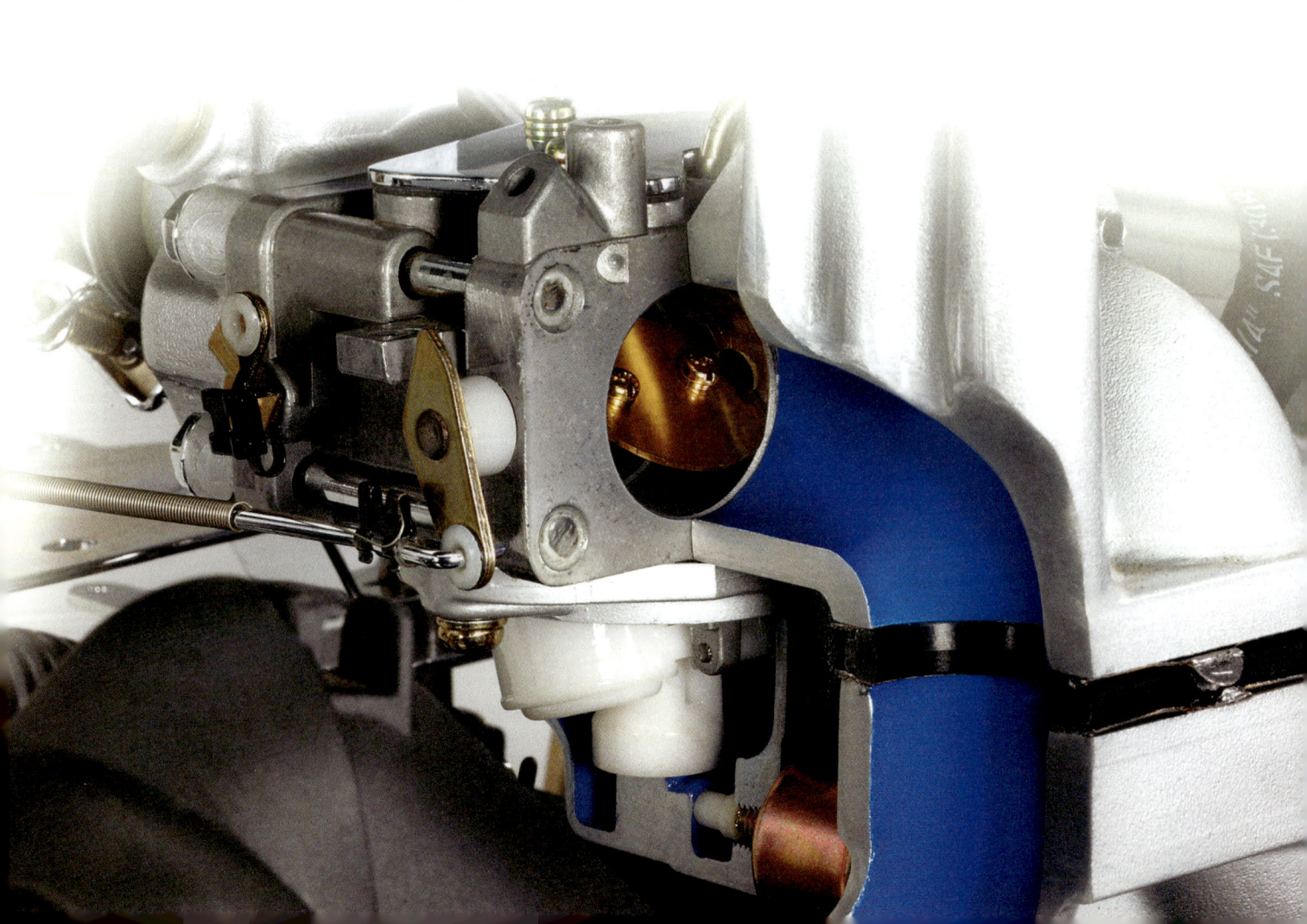

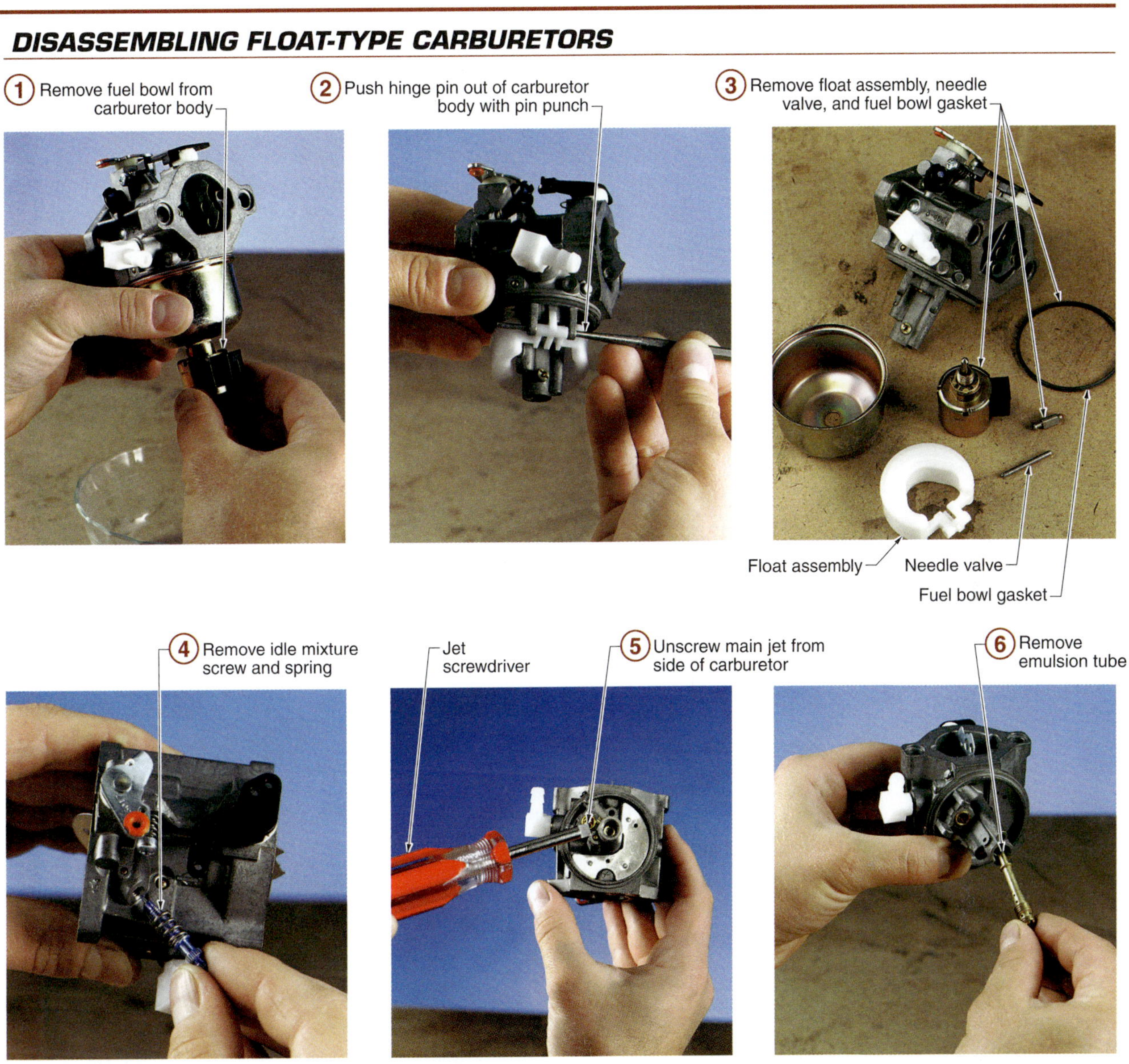

DISASSEMBLING FLOAT-TYPE CARBURETORS

1. Remove fuel bowl from carburetor body
2. Push hinge pin out of carburetor body with pin punch
3. Remove float assembly, needle valve, and fuel bowl gasket

Float assembly — Needle valve — Fuel bowl gasket

4. Remove idle mixture screw and spring
5. Unscrew main jet from side of carburetor — Jet screwdriver
6. Remove emulsion tube

Figure 7-3. A float-type carburetor can be disassembled with a few common hand tools.

...Project: Overhauling Carburetors

TECH FACT

The most well-designed engines of OPE, such as lawnmowers, lawn and garden tractors, and snow throwers, are equipped with automotive-type carburetors, which are mounted away from the fuel tanks with float bowls. The carburetors of chainsaws, leaf blowers, and other portable OPE are equipped with diaphragms to allow for better maneuverability.

Carburetor Overhaul

To clean and inspect a carburetor and idle mixture screw, apply the following procedure:

1. Soak carburetor parts in an all-purpose parts cleaning solvent for no more than 15 min to remove any contaminants, or spray with carburetor cleaner. *Note:* Always wear eye protection when using any carburetor cleaning solvent or using compressed air.
2. Wipe away solvent and other residue thoroughly, using a clean cloth. Never use wire or tools. They can damage or further obstruct openings.
3. Inspect all components and use additional carburetor cleaner to loosen stubborn deposits and to clear any obvious obstructions.
4. Carefully inspect the idle mixture screw for wear.
5. Replace an idle mixture screw if the tip is bent or contains a ridge. Replace any carburetor parts that are damaged or permanently clogged. ***See Figure 7-4.*** *Note:* An idle mixture screw is made of brass and controls the air-fuel mixture at high speeds and at idle. Overtightening the screw can damage the tip to the point that proper adjustment is no longer possible.

To reassemble and reattach a carburetor and air-cleaner assembly, apply the following procedure:

1. Use a carburetor repair kit to install an inlet needle seat, with the groove facing down, using a small pin punch.
2. Install the inlet needle on the float, and install the float assembly in the carburetor body.
3. Insert the hinge pin into the carburetor body.
4. Install the emulsion tube and main jet (if so equipped).
5. Install the idle mixture screw.
6. Install the rubber gasket on the carburetor and attach the fuel bowl, fiber washer, and bowl nut.
7. Position the carburetor so that the beveled edge fits into the fuel intake pipe. Then, attach the carburetor with nuts or bolts, as required, leaving these fasteners loose for final tightening with a torque wrench. *Note:* Consult an authorized Briggs & Stratton dealer for the proper amount of tightening torque.
8. Install the air-cleaner assembly, making certain that the tabs on the bottom of the air cleaner are engaged with the slots.

Chapter 7—Intermediate Small Engine Maintenance and Repair Projects 111

CLEANING CARBURETOR PARTS

① Soak carburetor parts in cleaning solvent

Idle mixture screw

② Wipe away residue with clean cloth

③ Reclean parts with stubborn deposits

Good tip

Damaged tip

④ Inspect idle mixture screw for wear or damage

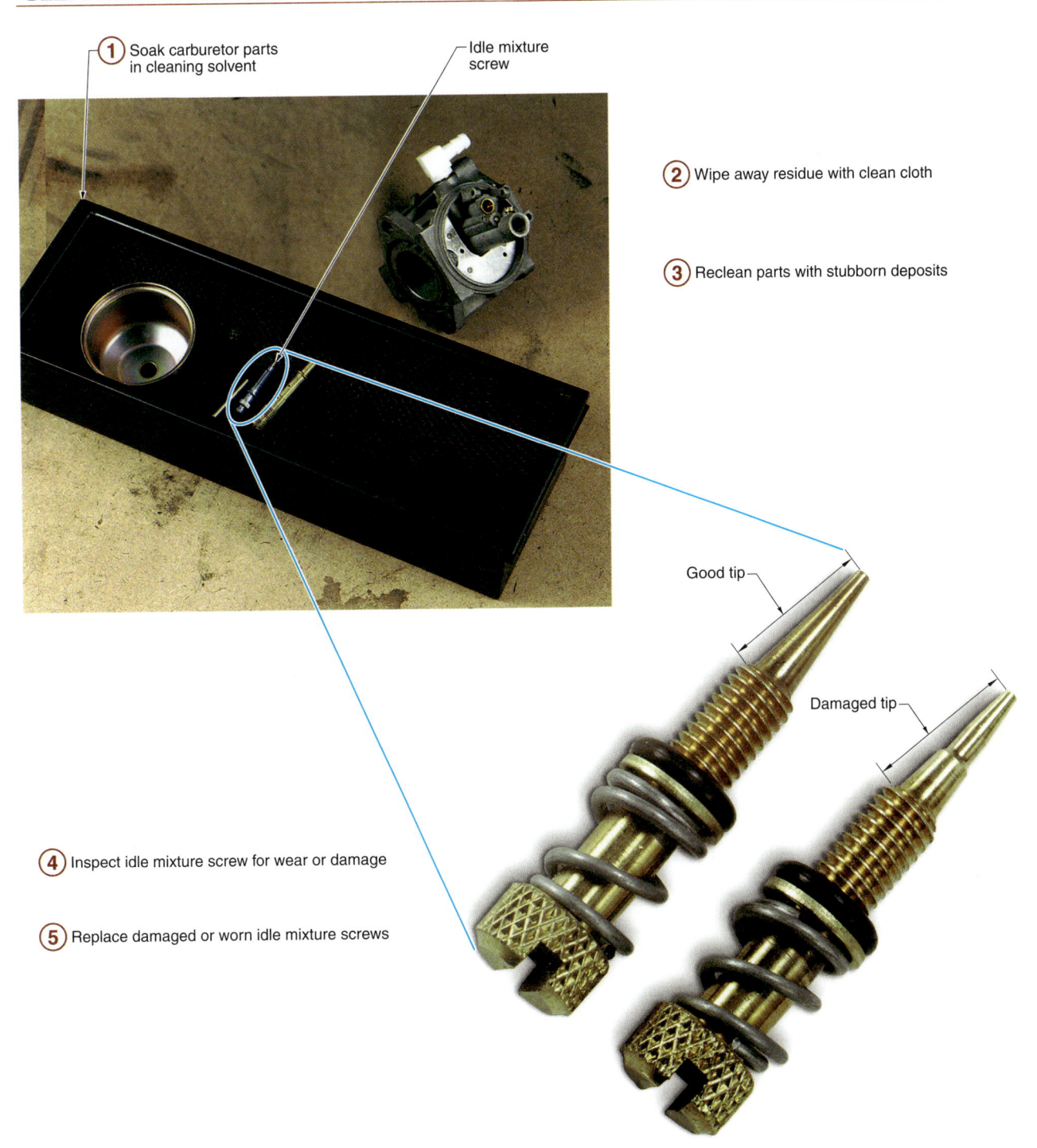

⑤ Replace damaged or worn idle mixture screws

Figure 7-4. Carburetor parts should be cleaned with carburetor cleaning solvent and inspected for wear or damage prior to reassembly.

Project: Replacing Flywheels and Flywheel Keys

- **Competency level:** intermediate
- **Tools needed:** socket wrench set, flywheel clutch tool, flywheel holder or strap wrench, flywheel puller, flat file, and torque wrench
- **Estimated completion time:** 60 min

The flywheel on a small engine was originally designed to store the momentum from combustion to keep the crankshaft turning between the power strokes of the engine. The flywheel on modern small engines serves several other purposes. The fins help to cool the engine by distributing air around the engine block. The fins also blow air across the air vane on a pneumatic governor, maintaining the desired engine speed. Magnets mounted on the outside surface of the flywheel are required for ignition.

If a lawn mower or tiller blade hits an immovable object, such as a rock, tree stump, or curb, the flywheel key can sometimes absorb the damage, significantly reducing repair costs. The key and keyway (key slot on the crankshaft) should be inspected for damage by removing the flywheel. The soft metal key must be able to eliminate play between the flywheel and crankshaft.

To replace a flywheel, apply the following procedure:

1. Disconnect the spark plug lead and secure it away from the spark plug. Remove the shroud by using a socket wrench to loosen the bolts holding the shroud in place. If the engine is equipped with a flywheel brake, proceed to Step 2. If the flywheel is equipped with a flywheel clutch, proceed to Step 3.
2. Remove any covers and disconnect the outer end of the brake spring.
3. Remove the flywheel clutch with a flywheel clutch tool while holding the flywheel with a flywheel holder or a flywheel strap wrench. If the flywheel is attached with a nut, use the flywheel holder as a brace, and remove the flywheel retaining nut with a socket wrench.
4. With the flywheel nut threaded onto the crankshaft, install a flywheel puller so that its bolts engage the holes adjacent to the hub of the flywheel. If the holes are not threaded, use a self-tapping flywheel puller, or tap the holes using a ¼ × 20 tap. **CAUTION:** Never strike a flywheel with a hand tool. Even a slightly damaged flywheel presents a safety hazard and must be replaced.
5. Rotate the puller nuts evenly until the flywheel pops free, and remove the flywheel and key.
6. Visually inspect the crankshaft and flywheel for cracks or broken fins and replace if damaged. *Note:* An undamaged flywheel has clean and smooth tapered sections, with no play between the sections.
7. Inspect the crankshaft and flywheel keyways for damage. Slight burrs may be removed with a flat file. There should be no play or wobbling when the flywheel is placed on the crankshaft.
8. Inspect the flywheel key and replace if there are any signs of shearing.
9. Place the flywheel on the crankshaft and look through the flywheel hub to align the keyways on the flywheel and crankshaft. Obtain a new flywheel key designed for the make and model of the OPE from an authorized service dealer.
10. Place the flywheel key in the keyway (it must fit securely).
11. Once the key and flywheel are securely in place, use a torque wrench to reattach the flywheel nut or clutch. Consult an authorized Briggs & Stratton dealer for the torque specifications of the OPE make and model. ***See Figure 7-5.***

Chapter 7—Intermediate Small Engine Maintenance and Repair Projects 113

REPLACING FLYWHEELS

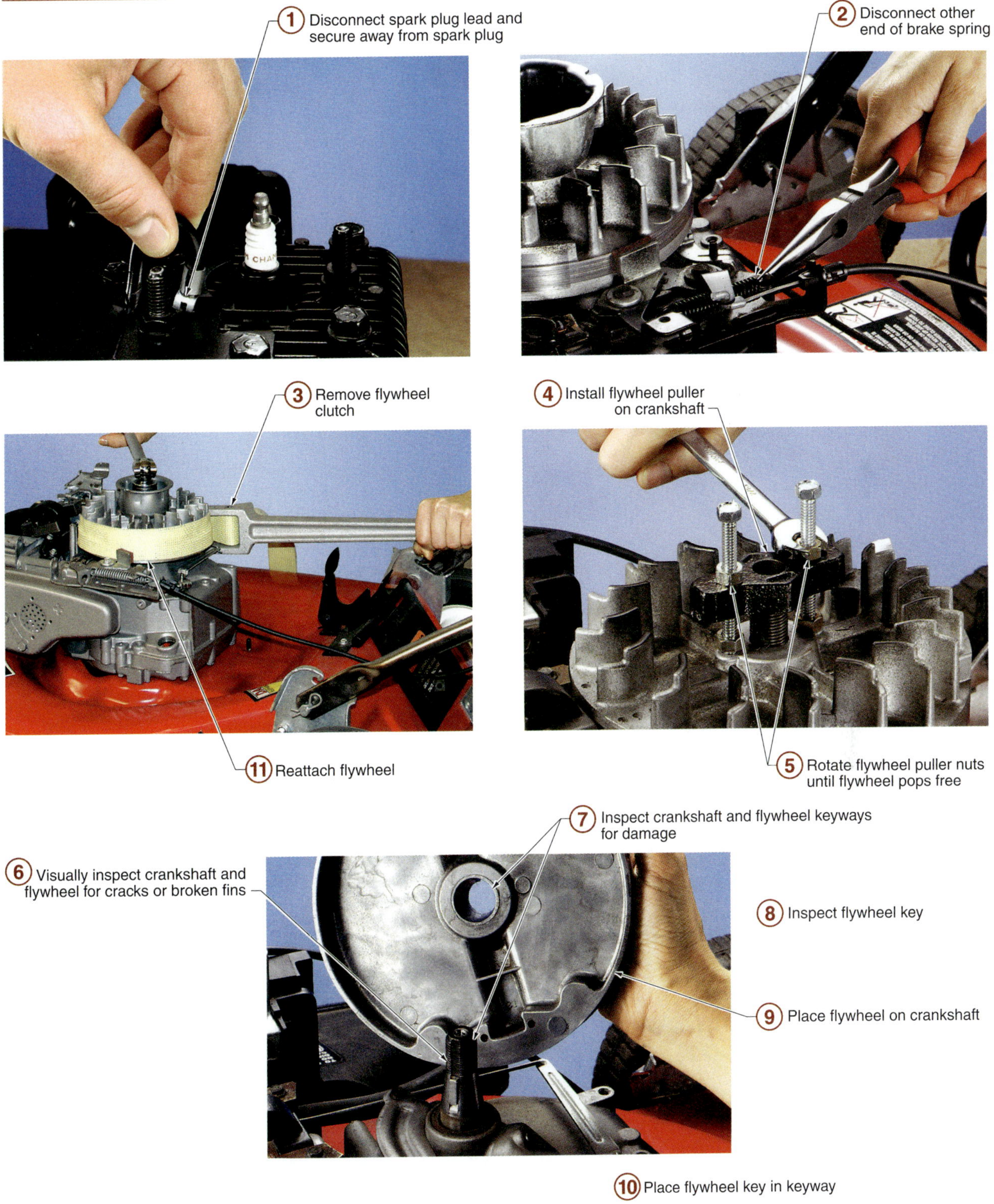

1. Disconnect spark plug lead and secure away from spark plug
2. Disconnect other end of brake spring
3. Remove flywheel clutch
4. Install flywheel puller on crankshaft
5. Rotate flywheel puller nuts until flywheel pops free
6. Visually inspect crankshaft and flywheel for cracks or broken fins
7. Inspect crankshaft and flywheel keyways for damage
8. Inspect flywheel key
9. Place flywheel on crankshaft
10. Place flywheel key in keyway
11. Reattach flywheel

Figure 7-5. The flywheel is removed to inspect for damage and to inspect the key and keyway.

Project: Servicing Flywheel Brakes . . .

- **Competency level:** intermediate
- **Tools needed:** needle-nose pliers, standard screwdrivers, ruler or caliper, digital multimeter (DMM), torque wrench, tang bending tool, standard pliers, and starter clutch adapter
- **Estimated completion time:** 60 min

Most small rotary lawn mowers are equipped with an operator-presence safety control lever commonly referred to as a brake bail. Within 3 sec of being released, the brake bail stops a running engine and any cutting blade directly attached to the crankshaft of the engine. The stop switch immediately grounds the ignition, shutting off the engine, while a brake pad or band stops the flywheel from spinning. If the engine operates for more than 3 sec after the bail is released, the stop switch may be faulty. If the blade rotates for more than 3 sec after the bail is released, the brake pad or band may be in need of adjustment or be worn and need to be replaced.

Some older models are equipped with a brake band that may require adjustment by an authorized Briggs & Stratton service technician. Servicing a flywheel brake involves the inspection and testing of the brake pad system and brake band system.

Inspecting and Testing Brake Pad Systems

To inspect and test a brake system, the brake pad is first removed. The brake pad is then inspected for nicks, cuts, debris, and other damage. The stop switch is tested with a digital multimeter (DMM) before the braking system is reassembled.

To remove a brake pad, apply the following procedure:

1. Remove the spark plug lead and secure it away from the spark plug. Remove any other components that block access to the brake, such as the finger guard, fuel tank, oil fill tube, blower housing, or rewind starter.
2. Remove the brake-control bracket cover, if equipped. Loosen the cable clamp screw with a screwdriver, and remove the brake cable from the control lever.
3. Use needle-nose pliers to disconnect the spring from the brake anchor.
4. Remove the stop switch wire from the stop switch by gently squeezing the switch and lightly pulling on the wire until it slips free. If the engine is equipped with an electric starter motor, disconnect the pair of wires leading to the starter motor.
5. Loosen the brake bracket screws with the screwdriver, and remove the bracket from the brake assembly. *See Figure 7-6.*

To inspect and test a brake pad system, apply the following procedure:

1. Visually inspect the brake pad for nicks, cuts, debris, and other damage. Check for wear by measuring the thickness of the pad with a ruler or caliper. *Note:* Measure the pad only, not the bracket. Replace the brake assembly if the thickness of the pad is less than 0.090″.
2. Test the stop switch by using a DMM set to measure continuity to determine whether the ignition circuit is grounded when the stop switch is activated. The stop switch should indicate continuity (0 Ω) to engine ground when the switch is set to STOP and no continuity (∞) when the switch is set to RUN.
3. If a problem exists, check for loose or faulty connections. *See Figure 7-7.*

REMOVING BRAKE PADS

① Remove spark plug lead and secure away from spark plug

② Remove brake control bracket cover, oil fill tube, and blower housing

⑤ Remove bracket from brake assembly

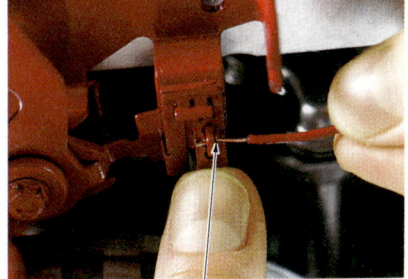

③ Disconnect spring from brake anchor

④ Remove stop switch wire from stop switch

Figure 7-6. If a lawn mower blade rotates for more than 3 sec after the brake bail is released, the brake pad or band may be worn or in need of adjustment.

INSPECTING AND TESTING BRAKE PAD SYSTEMS

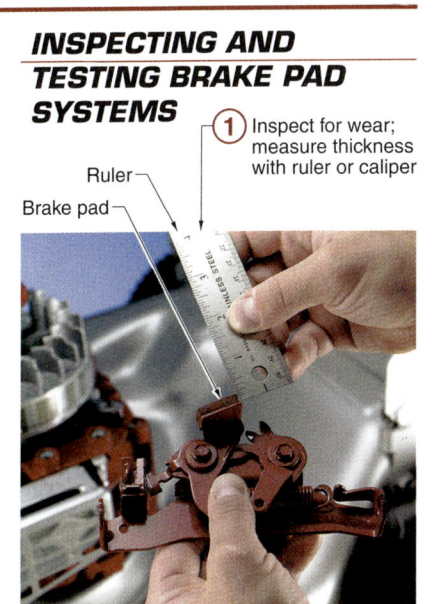

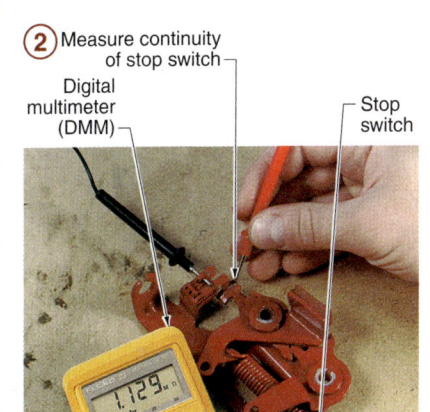

① Inspect for wear; measure thickness with ruler or caliper

Ruler

Brake pad

② Measure continuity of stop switch

Digital multimeter (DMM)

Stop switch

③ Check for loose or faulty connections

Figure 7-7. Brake pads are inspected for wear, measured for thickness, and tested for continuity after removal from a small engine.

116 SMALL ENGINE AND EQUIPMENT MAINTENANCE

... Project: Servicing Flywheel Brakes ...

To reassemble a brake pad system, apply the following procedure:

1. Install the brake assembly on the cylinder and tighten the mounting bolts to 40 in-lb, by using a torque wrench.
2. Install the stop switch wire, bending the end of the wire to a 90° angle.
3. Install the blower housing and any other engine components removed while servicing the brake system.
4. Check the braking action by pivoting the lever. Verify that the lever moves freely and the pad makes full contact with the flywheel.
5. Attach the brake spring using needle-nose pliers, and connect the brake cable that connects to the brake bail.
6. Test the brake pad system by starting the engine and then releasing the brake bail. The engine and the blade or other equipment should come to a complete stop within 3 sec. *Note:* If there is uncertainty regarding the effectiveness of the brake pad system, consult an authorized Briggs & Stratton service technician for further inspection. *See Figure 7-8.*

REASSEMBLING BRAKE SYSTEMS

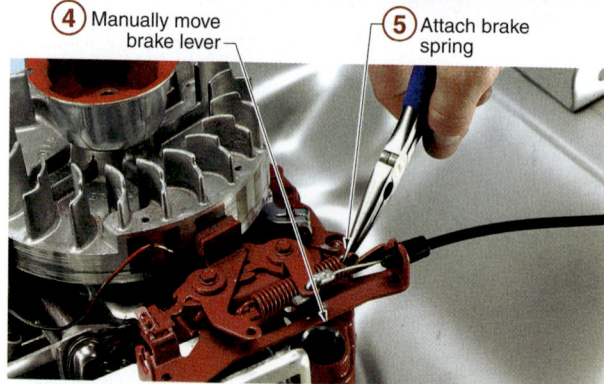

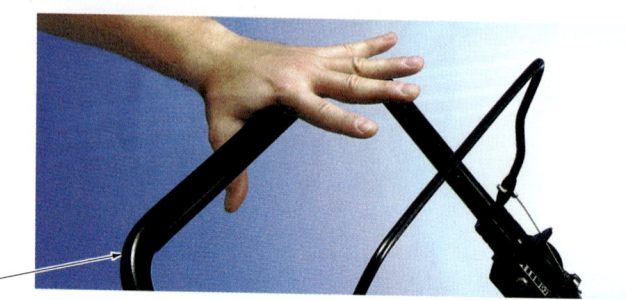

Figure 7-8. A brake system can be reassembled in six steps.

Inspecting and Testing Brake Bands

The only safe method of using a brake bail on a small engine is pulling and holding the bail by hand when starting and running the engine. It can be released when necessary to stop the engine. Keeping the bail in the operating position by any other means overrides an important safety mechanism. The bail is required by law and designed to protect the operator from injury. The brake band contains loops at both ends and is mounted on a stationary post and a movable post. A tang over the movable post prevents the brake band from dislodging during operation.

To remove and inspect a brake band, apply the following procedure:

1. Use a tang bending tool to bend the control lever tang outward so that it clears the brake band loop.
2. Use standard pliers to release the brake spring.
3. Lift the band off the stationary and movable posts.
4. Inspect the band for damage. Replace it if nicks or cuts are present.
5. Check for wear by measuring the thickness of the brake pad with a ruler or caliper. *Note:* Measure the pad only (not the metal band). Replace the brake band if the thickness of the pad is less than 0.030″. **See Figure 7-9.**

REMOVING AND INSPECTING BRAKE BANDS

① Use tang bending tool to bend control lever tang outward

② Release brake spring

③ Lift band off stationary and movable posts

④ Inspect band for damage

⑤ Measure band thickness with caliper or ruler

Figure 7-9. A brake band has loops at both ends. It is mounted on a stationary and a movable post. A tang over the movable post prevents the brake band from dislodging during operation.

...Project: Servicing Flywheel Brakes

To reassemble a brake band, apply the following procedure:

1. Reinstall the stop switch wire on the control bracket. On older systems, reinstall the stop switch wire on the control bracket of the stop switch terminal.
2. Place the brake band on the stationary post and loop it over the end of the movable post until the band bottoms out. *Note:* The brake material on a steel band must be on the flywheel side after assembly. On older systems, install the brake band on the stationary and movable posts.
3. Bend the retainer tang until it is positioned over the brake band loop so that the loop cannot be accidentally dislodged. After assembly, check that the braking material on the metal band faces the flywheel.

To test a brake band, apply the following procedure:

1. Remove the spark plug lead and secure it away from the spark plug. For an electric start engine, disconnect and remove the battery.
2. With the brake engaged, turn the starter clutch, using a starter clutch adapter and torque wrench. Turning the flywheel clockwise at a steady rate should require at least 45 in-lb of torque. If the torque reading is lower than 45 in-lb, the components may be worn, damaged, or in need of adjustment.
3. Test the stop switch with a DMM set to measure continuity to determine whether the ignition circuit is grounded when the stop switch is activated. The stop switch should indicate continuity (0 Ω) to engine ground when the switch is set to STOP and no continuity (∞) when the switch is set to RUN.
4. If a problem exists, check for loose or faulty connections. ***See Figure 7-10.***

TESTING BRAKE BANDS

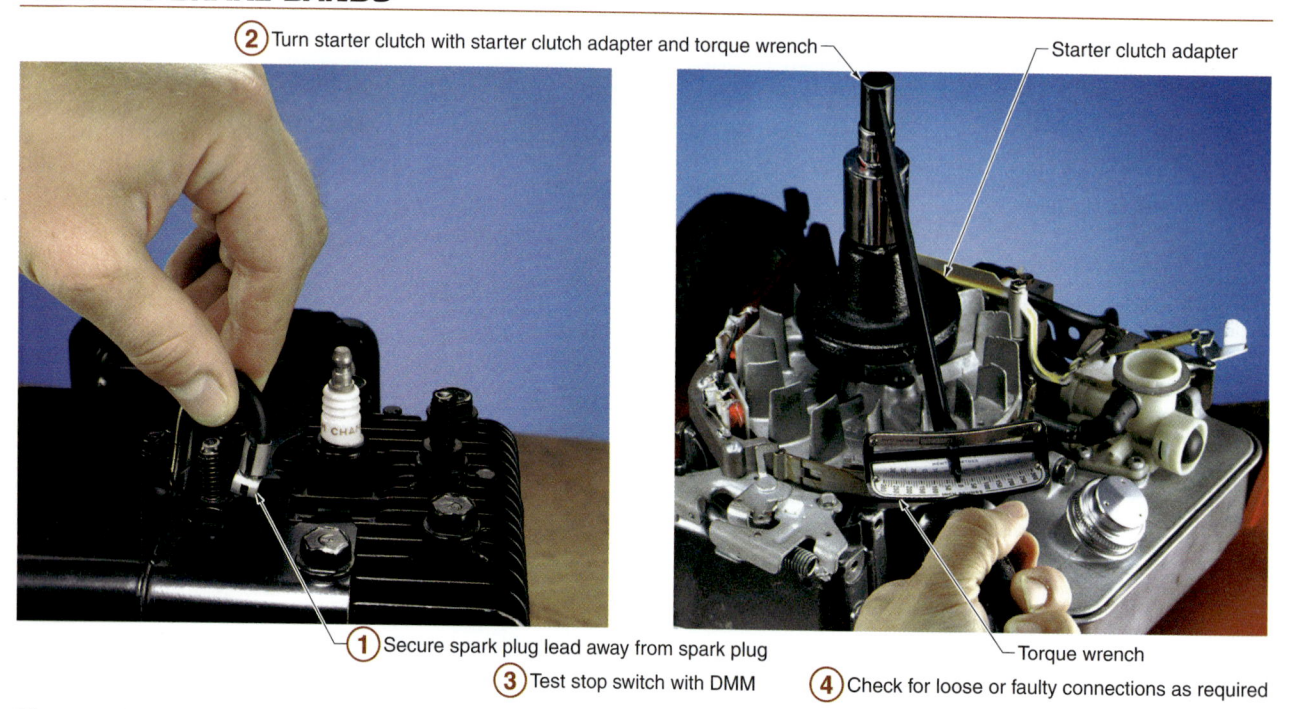

Figure 7-10. A band brake is tested with a starter clutch adapter and torque wrench.

Project: Servicing Fuel Systems...

- **Competency level:** intermediate
- **Tools needed:** flashlight, baster, fuel line clamp, standard screwdriver, needle-nose pliers, torque wrench, carburetor cleaner, and tachometer
- **Estimated completion time:** 30 min

Fuel tanks are designed to keep engine fuel clean, vented, and secure. Maintenance is needed if there is debris in the tank or the tank leaks fuel. A fuel tank with a crack or hole must be replaced. Older fuel tanks are made of steel, while modern models are made of steel or polymeric material. Regardless of the tank material, a damaged fuel tank must not be repaired but should be replaced. Tanks are typically installed as far away as possible from the hottest areas of the engine to keep the fuel cool.

When replacing a fuel tank, only parts recommended by the engine original equipment manager (OEM) should be used. These parts will attach securely to an engine in the space provided. Many fuel tanks are designed to use a vented fuel cap to prevent a vacuum from forming in the fuel line. If fuel is leaking from the cap, a properly fitted replacement cap can be used to stop the leakage. Servicing a fuel system involves cleaning the tank; inspecting the fuel filter; servicing the fuel pump; and adjusting the carburetor, a number of mixture screws, and the choke linkage.

Cleaning Fuel Tanks

As fuel is added into the fuel tank of outdoor power equipment throughout the season, small particles of dirt, dust, and other debris can possibly fall into the fuel tank. Over time, this debris can accumulate and block the opening to the fuel line. Therefore, fuel tanks must be inspected for damage and cleaned.

To clean a fuel tank, apply the following procedure:

1. Remove the spark plug lead and secure it away from the spark plug.
2. Check inside the tank with a flashlight for debris or other contaminants. Use a baster to remove loose debris.
3. Use a fuel line clamp or other smooth-face clamp to seal the fuel line where it attaches to the carburetor; remove the fuel tank from the engine.
4. Disconnect the line from the carburetor, hold the line over a bucket or fuel can, and release the clamp. Dispose of all fuel in a safe manner.
5. Inspect the fuel filter for debris or deposits.
6. Reattach the fuel tank or install a new tank. Use a screwdriver to fasten it firmly with cap screws. *Note:* This is a good opportunity to replace the fuel line and filter using OEM-approved replacement parts. ***See Figure 7-11.***

CLEANING FUEL TANKS

① Remove spark plug lead and secure away from spark plug

② Inspect inside of fuel tank; use baster to remove excess debris from tank

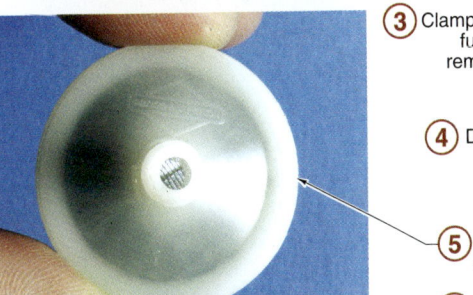

③ Clamp fuel line with fuel line clamp; remove fuel tank from engine

④ Disconnect fuel line from carburetor and allow to drain into bucket

⑤ Inspect fuel filter for debris or deposits

⑥ Reattach fuel tank

Figure 7-11. If debris can be seen in the tank or the tank is leaking fuel, the fuel tank must be removed and cleaned.

120 SMALL ENGINE AND EQUIPMENT MAINTENANCE

. . . Project: Servicing Fuel Systems . . .

Fuel tanks must be constructed of a noncorrosive material or coated with a corrosion-resistant layer to protect against the damaging effects of water, alcohol, and salt. If the tank is designed to deliver fuel through a fuel line, a convex fuel filter may be located at the base of the tank, where fuel from the tank enters the fuel line. A fuel filter can also be located outside the tank, midway along the fuel line. *See Figure 7-12.* If a fuel tank must sustain excessive vibration, a labyrinth-equipped tank can be installed on some models. A labyrinth tank, available from an authorized Briggs & Stratton dealer, contains a set of baffles and/or a foam insert to reduce the sloshing and vaporization of fuel.

Inspecting Fuel Lines

Inspecting Fuel Filters

A clean fuel filter strains fuel before it reaches the carburetor and prevents foreign particles from clogging the engine. A dirty fuel filter can cause the engine to run too lean, with diminished performance and uneven operation. Other factors can cause these problems, but the fuel supply is simple to check. Filters contain either a mesh screen or a pleated paper element. Some fuel filters are located inside the tank, while others are fitted into the fuel line between the tank and the fuel pump.

Fuel filters are rated by the size of the holes in the filtering material, in microns (μm). The size of the holes in the filter determines the largest particles that can flow through the filter. The number of holes affects the amount of fuel that can flow through the filter. Briggs & Stratton mesh screen filters are color-coded: red for 150μm and white for 75μm. *See Figure 7-13.* Pleated paper filters, designed for use in the fuel tanks, are typically contained in a clear plastic casing and rated 60μm. They consist of multiple folds that strain out particles suspended in the fuel. The proper filter for an engine depends on the engine design. An authorized Briggs & Stratton dealer should be consulted for the correct replacement filter.

FUEL FILTER COLOR-CODING

150μm (RED)

75μm (WHITE)

Figure 7-13. Briggs & Stratton mesh screen filters are color-coded red for 150μm and white for 75μm.

FUEL FILTERS

Fuel tank — Fuel filter — Fuel line

Figure 7-12. A fuel filter can be located outside the tank, midway along the fuel line.

To inspect a fuel filter, apply the following procedure:

1. Close the fuel shutoff valve located at the base of the fuel tank where the fuel line is attached (if equipped). If the tank is not equipped with a fuel valve, clamp the fuel line with a fuel line clamp. Use a dry cloth to hold the filter and catch any dripping fuel.
2. If the filter is installed inside the tank, drain the tank before removing the filter for inspection or replacement. If the filter is not in the tank, proceed to Step 3.
3. If the filter is installed in the fuel line, remove the metal clips on each side of the filter, using needle-nose pliers, and slide the filter out of the fuel line.
4. Shake the filter over a clean cloth to displace any remaining fuel. Use the cloth to wipe off any residue from the outside of the filter. Keep the filter a safe distance away from the face and view through one end. With a clean filter, there should be light shining through clearly from the opposite side. If debris is clogging the mesh screen, pleated paper, or inside of the casing, replace the filter. ***See Figure 7-14.***

Note: Always wear safety glasses or goggles when removing or inspecting a filter to protect the eyes from liquid fuel or fuel vapors.

INSPECTING FUEL FILTERS

① Close fuel shutoff valve

② Drain tank of fuel

③ Slide filter out of fuel line

④ Inspect fuel filter for debris or dirt

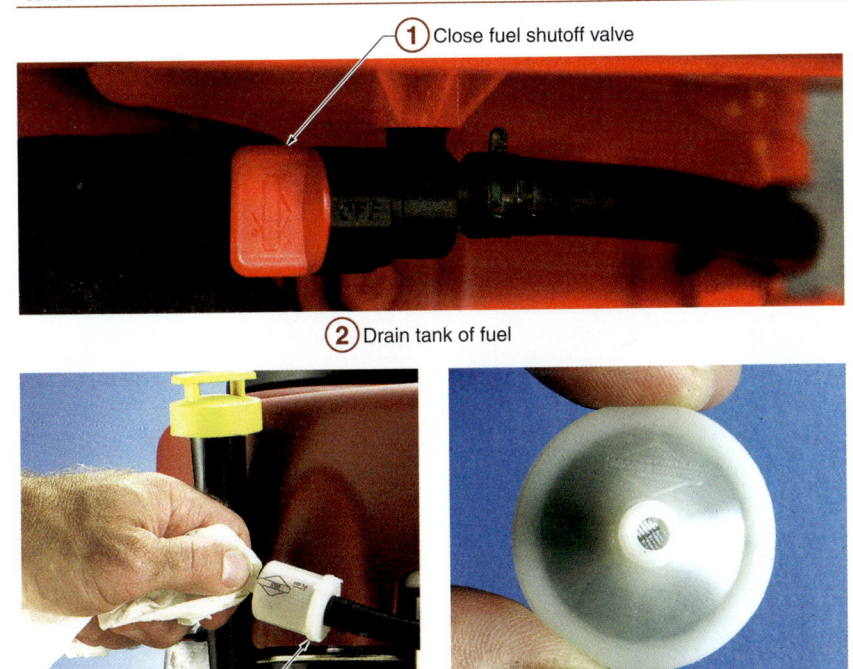

Figure 7-14. Fuel filters should be inspected periodically for the presence of dirt and other small particles.

Gasoline Use

Only fresh unleaded gasoline should be used in a small engine. Other considerations for gasoline use in four-cylinder small engines include the following:

- Use fresh, clean unleaded gasoline with a minimum octane of 87 for both L-head and overhead valve engines. Since small engines operate at relatively low compression ratios, knocking is seldom a problem, so using gasoline with a higher octane rating is unlikely to offer any benefit.
- Gasoline that is over a month old and does not contain a gasoline stabilizer should not be used in an engine. Hard starting and varnish formation may result from its use.
- Drain the tank if fuel sits for more than a month. The Environmental Protection Agency (EPA) recommends pouring old fuel into an automobile gas tank. As long as the tank is at least half full, the old fuel will mix harmlessly with the new and will not affect its engine.
- Four-stroke cycle engines are used on most lawn mowers and large lawn equipment. Never use an oil-gasoline mixture in a four-stroke cycle engine, since the engine has an independent oil supply. A four-stroke cycle engine has an oil-fill cap leading to the crankcase.

...Project: Servicing Fuel Systems...

Servicing Fuel Pumps

A fuel pump is used when the fuel tank is mounted lower than the carburetor and gravity cannot be relied upon to carry fuel through the fuel line. Briggs & Stratton fuel pumps have either a plastic or a metal body and develop pressure using the vacuum in the crankcase, which is created by the motion of the piston. A fitting on the crankcase cover or the dipstick tube draws on the crankcase vacuum to create the pressure to pump fuel. The fuel pump may be mounted on the carburetor, near the fuel tank, or between the tank and carburetor.

To service a fuel pump, apply the following procedure:

1. Close the fuel shutoff valve (if equipped) at the base of the fuel tank, where the fuel line is attached. If there is no fuel valve, stop the flow of fuel using a fuel line clamp.
2. Loosen the mounting screws and remove the fuel pump from the mounting bracket or carburetor.
3. Inspect the pump for hairline cracks and other damage to the external surfaces of the pump. If the pump is damaged and has a metal body, discard the pump and install a replacement pump from the OEM. A plastic pump can be serviced by using an OEM repair kit to replace worn parts. A damaged metal pump must be replaced.
4. With the fuel valve closed or the line clamped, remove the mounting screws.
5. Disconnect the fuel hose using needle-nose pliers to loosen the clips.
6. Remove the screws and disassemble the pump.
7. Inspect the pump body for cracks or other damage. Soak metal parts in an all-purpose parts cleaning solvent. The pump body may be soaked for up to 15 min.
8. Inspect the hose for cracks and softening or hardening. Replace any faulty parts. Discard old gaskets, diaphragms, and springs, and replace them with parts from an OEM repair kit.
9. Place the diaphragm spring and then the cup over the center of the pump chamber. Insert the valve spring.
10. Install the diaphragm, gasket, and cover, and then attach with pump screws.
11. Tighten the screws to between 10 in-lb and 15 in-lb using a torque wrench.
12. Reattach the pump to the carburetor or mounting bracket using the pump mounting screws. *See Figure 7-15.*

SERVICING FUEL PUMPS

① Close fuel shutoff valve

② Remove fuel pump from mounting bracket — Pump

③ Inspect fuel pump for damage — Crack in pump body

⑧ Inspect fuel line hose

⑫ Reattach pump

④ Remove fuel pump mounting screws

⑤ Disconnect fuel hose

⑥ Disassemble fuel pump

⑦ Soak metal parts in parts cleaning solvent

⑨ Place diaphragm spring and cup over pump body

Pump screws

Diaphragm spring

Cup

Diaphragm

Cover

Gasket

Pump body

⑩ Install diaphragm, gasket, and cover

⑪ Install and tighten pump screws

Figure 7-15. Fuel pumps must be periodically serviced due to ordinary use or to check for possible damage.

...Project: Servicing Fuel Systems...

Cleaning and Checking for Fuel Flow from Tank to Carburetor

When an engine does not start or performs poorly, there may be a problem with the carburetor. The carburetor of a small engine eventually accumulates deposits by encountering grass, twigs, and other debris. Deposits inside the carburetor can clog fuel and air passages. As a result, engine performance may be reduced or the engine may stop. Fortunately, many of the problems associated with the accumulation of deposits can be addressed quickly and easily, often without even removing the carburetor from the engine. For example, a carburetor cleaner, which is available in convenient spray cans, may be used to periodically clean both inside and outside the carburetor. The source of the problem may also be in the fuel valve, filter, or pump.

For smooth engine operation, it is also important to keep the carburetor and linkages clean and properly adjusted. The linkages attached to the throttle and choke plates of the carburetor can bind or stick when dirty. Constant vibration and wear can affect the settings of the mixture screws.

To clean a carburetor, apply the following procedure:

1. Remove the air cleaner and inspect the choke plate mounted on the shaft at the opening of the carburetor throat. Ensure that the choke plate closes easily and completely. A choke that does not move freely or close properly can cause difficulties in starting.
2. Spray a small amount of carburetor cleaner on the shaft of a sluggish choke and into the venturi to loosen grit. Debris in the carburetor often causes performance problems.
3. Open the fuel shutoff valve (if equipped), located at the base of the fuel tank where the fuel line is attached. Remove the fuel line and check for blockage. Fuel cannot reach the carburetor if the fuel valve is closed.
4. Remove and inspect the spark plug. A wet spark plug may indicate overchoking, the presence of water in the fuel, or an excessively rich fuel mixture. A dry spark plug may indicate a plugged fuel filter, leaking mounting gaskets on either end of the carburetor, or a stuck or clogged carburetor inlet needle.
5. Pour a teaspoon of fuel into the spark plug hole.
6. Screw the spark plug back in, reattach the spark plug lead, and start the engine. If the engine fires only a few times and then quits, assume a dry plug condition and consider the causes of a dry plug, as listed in Step 4. *See Figure 7-16.*

Chapter 7—Intermediate Small Engine Maintenance and Repair Projects 125

ADJUSTING CARBURETORS

② Spray carburetor cleaner on choke

③ Open fuel shutoff valve

① Remove air cleaner and inspect choke plate

⑤ Pour 1 tsp of fuel into spark plug hole

④ Remove and inspect spark plug

⑥ Screw in spark plug and start engine

Figure 7-16. Proper fuel system service requires adjusting the carburetor.

SMALL ENGINE AND EQUIPMENT MAINTENANCE

...Project: Servicing Fuel Systems...

Adjusting Idle Speed Screws and Idle Mixture Screws

On some float-type carburetors, the air-fuel mixture and engine speed can be adjusted at idle. Fuel system service involves inspecting and adjusting the idle speed screw and idle mixture screw. The idle speed screw keeps the throttle plate from closing completely. The idle mixture screw limits the flow of fuel at idle.

To adjust an idle speed screw and idle mixture screw, apply the following procedure:

1. With the engine off, remove the air filter and air cartridge.
2. Locate the idle mixture screw and turn it clockwise until the needle lightly touches the seat. Then, turn the screw counterclockwise 1½ turns.
3. If the carburetor has a main jet adjustment screw at the base of the float bowl, turn the screw clockwise until it just touches the seat inside the emulsion tube. Then, turn the screw counterclockwise 1 turn to 1½ turns. Replace the air cleaner and start the engine for final carburetor adjustments.
4. Run the engine for 5 min at half throttle to bring it to its operating temperature.
5. Slowly turn the idle mixture screw clockwise until the engine begins to slow. Turn the screw in the opposite direction until the engine again begins to slow.
6. Turn the screw back to the midpoint.
7. Using a tachometer to gauge engine speed, set the idle speed screw to bring the engine to 1750 rpm for aluminum-cylinder engines or 1200 rpm for engines with a cast iron cylinder sleeve.
8. Run the engine at idle (turtle icon).
9. Hold the throttle lever against the idle speed screw to bring the engine speed to "true idle." Then, repeat the idle mixture screw adjustments from Step 4 to fine-tune the mixture. *See Figure 7-17.*

ADJUSTING IDLE SPEED AND IDLE MIXTURE SCREWS

① Remove air filter and air cartridge

② Turn idle mixture screw clockwise until it touches seat; turn screw back 1½ turns

③ Adjust jet adjustment screw at base of bowl as required

④ Run engine for 5 min

⑤ Turn idle mixture screw clockwise until engine slows; reverse until engine again slows

⑥ Turn idle mixture screw to midpoint

⑦ Use tachometer to gauge engine speed

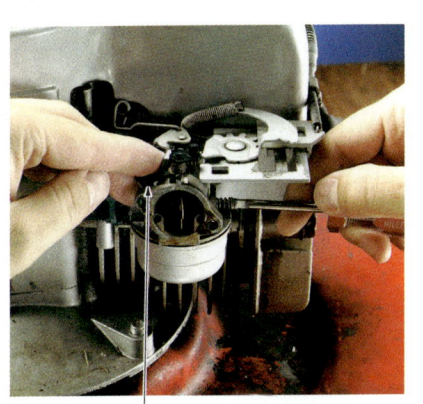

⑧ Run engine at idle

⑨ Hold throttle lever against idle speed screw

Figure 7-17. An engine can be adjusted to true idle by removing the air-cleaner assembly and rotating the idle speed mixture screw while the engine is in operation.

...Project: Servicing Fuel Systems

Adjusting High-Speed Mixture Screws

Some older carburetors contain a high-speed mixture screw. It is located near the throttle plate and opposite the idle speed screw. Under load, the high-speed circuit increases airflow through the throat. Setting the high-speed mixture involves running the engine until it is warm, stopping it to adjust the high-speed mixture, and then restarting for final adjustments. *Note:* On engines featuring both an idle mixture screw and a high-speed screw, the idle mixture adjustment procedure must be completed before attempting to adjust the high-speed mixture.

To adjust a high-speed mixture screw, apply the following procedure:

1. Run the engine for 5 min at half throttle to bring it to its operating temperature.
2. Stop the engine.
3. Locate the high-speed mixture screw, and turn it clockwise until the needle just touches the seat.
4. Turn the screw counterclockwise 1¼ turns to 1½ turns.
5. Restart the engine and set the throttle to the high position (rabbit icon).
6. Turn the high-speed or main jet screw clockwise until the engine begins to slow.
7. Turn the screw back to the midpoint.
8. Once adjusted, check the engine acceleration by moving the throttle from idle (turtle icon) to high (rabbit icon). The engine should accelerate smoothly. If necessary, readjust the mixture screws. **See Figure 7-18.**

ADJUSTING HIGH-SPEED FUEL MIXTURES

① Run engine for 5 min
② Stop engine
③ Turn high-speed mixture screw until it touches seat
④ Back high-speed mixture screw off 1¼ to 1½ turns
⑤ Restart engine and set throttle to high
⑥ Turn high-speed screw clockwise until engine begins to slow; reverse direction until engine again slows
⑦ Turn screw back to midpoint
⑧ Move throttle lever from idle to high; adjust mixture screws as required

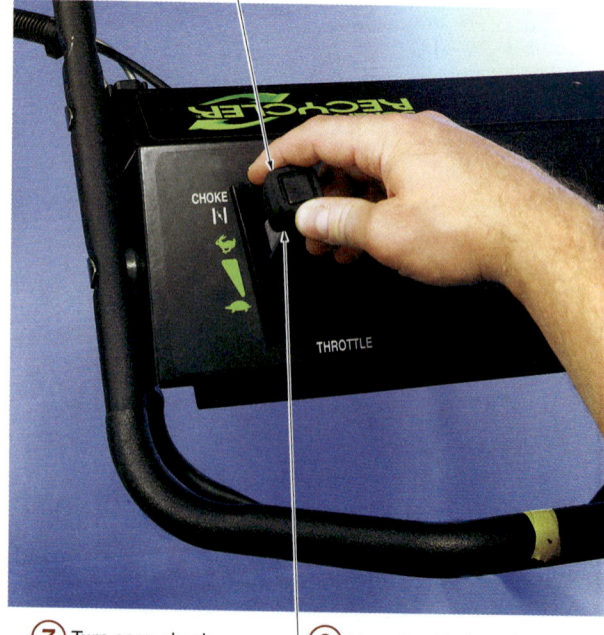

Figure 7-18. The high-speed fuel mixture on an engine is adjusted by rotating the high-speed mixture screw and moving the throttle lever.

Adjusting Choke Linkages

Over time, choke linkages can become loose from normal operation, causing rough engine operation. Loose choke linkage can be easily corrected by using a simple adjustment procedure.

To adjust choke linkage, apply the following procedure:

1. Remove the air cleaner, and locate the choke lever on the engine or remote engine speed controls. Remove the spark plug lead from the spark plug and secure it safely away from the engine.
2. Move the throttle control to the high position (rabbit icon).
3. Loosen the cable mounting bracket screw to allow movement of the cable casing.
4. Move the cable casing so that the choke is closed.
5. Tighten the bracket screw and check the motion of the control lever. Repeat steps, as necessary, until the cable moves freely. *See Figure 7-19.*

ADJUSTING CHOKE LINKAGES

① Remove air cleaner
② Move throttle to high position
③ Loosen cable mounting bracket screw
⑤ Tighten mounting bracket screw
④ Move cable casing so that choke is closed

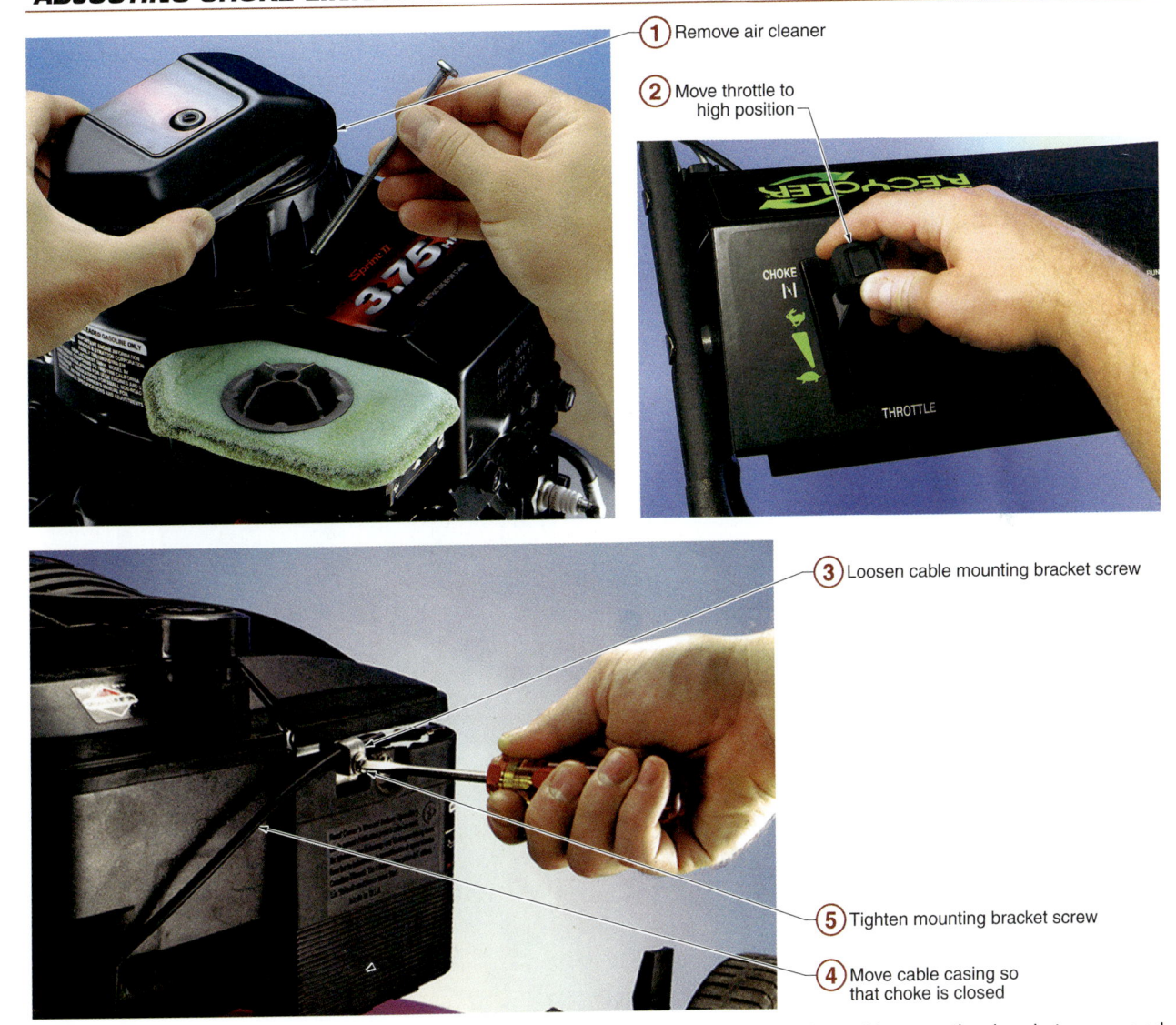

Figure 7-19. The choke linkage on a small engine is easily adjusted by loosening the cable mounting bracket screw and moving the cable casing until the choke is closed.

Project: Servicing Governor Systems...

- **Competency level:** intermediate
- **Tools needed:** standard screwdrivers, lubricating oil, socket wrench set or nut driver set, tang bending tool, and tachometer
- **Estimated completion time:** 30 min

A properly adjusted governor can maintain a steady engine speed regardless of changes in the terrain and other conditions that increase the work of the engine. These conditions are known as the "load." When engine speed starts to rise or fall in response to a change in the load, the governor responds by opening or closing the throttle. If engine speed is adjusted manually, using the equipment controls, the governor maintains the new setting.

An engine contains either a pneumatic governor or a mechanical governor. The blower housing must be removed to determine which type of governor is installed. The linkages of a pneumatic governor connect to the pivoting air vane next to the flywheel. The linkages of a mechanical governor connect to the governor shaft.

Note: Governor adjustment procedures vary widely depending on the make and model of the engine. Check with an authorized Briggs & Stratton dealer for the proper speed settings.

Mechanical Governor Components

A *mechanical governor system* is an engine system consisting of a gear assembly that meshes with a camshaft or other engine components to sense and maintain a desired engine speed. Engine speed is sensed by a gear-driven component that meshes with a drive gear and rotates at approximately crankshaft speed. A mechanical governor system is the most common type of governor system used on small air-cooled engines built in the last decade. *See Figure 7-20.* A mechanical governor system consists of the following components:

MECHANICAL GOVERNOR COMPONENTS

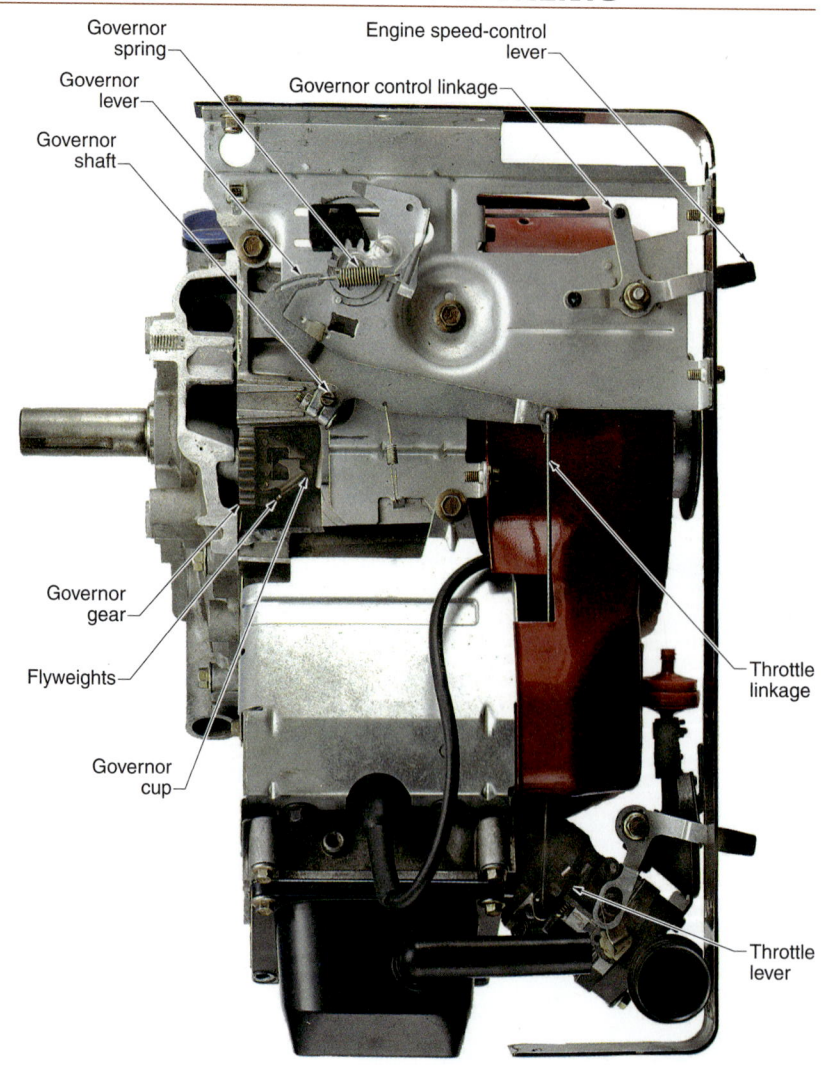

Figure 7-20. A mechanical governor system is the most common type of governor system used on modern small air-cooled engines.

- **engine speed-control lever:** Moving this lever to a higher speed setting opens the throttle indirectly by pulling on the governor gear bracket.
- **governor control linkage:** The bracket pivots to increase tension on the governor spring.
- **governor spring:** Tension on the governor spring pulls on the governor lever in an effort to open the throttle plate.
- **governor lever:** Its pivoting action pulls on the throttle linkage, applying pressure to the governor shaft.
- **governor shaft:** The governor shaft links the governor linkage and levers to the governor cup and other parts inside the crankcase.
- **throttle linkage:** Governor linkage pulls the throttle lever.
- **throttle lever:** The throttle lever opens the throttle plate, allowing more air-fuel mixture into the combustion chamber and causing the engine speed to increase.
- **governor gear:** Increased engine speed causes the governor gear to rotate faster and flyweights to move outward.
- **flyweight:** Movement of the flyweights applies pressure to the governor cup.
- **governor cup:** The governor cup allows the governor lever to pivot.

Mechanical Governor Systems

Throttle/Governor Assemblies

To service a mechanical governor system, apply the following procedure:

1. Disconnect the spark plug lead and secure it away from the spark plug.
2. Check that the governor linkages are attached and move freely by pulling gently on the throttle lever. This should stretch the governor spring while pressing on the governor lever. If not, check that the governor spring and the link to the governor lever are properly attached to the throttle lever. *Note:* Springs and linkages that are not attached may be reconnected if they are in good condition. Twist them carefully into place to ensure that the delicate springs and linkages are not permanently bent or stretched. Do not use pliers or other tools to bend or distort links or springs.
3. Replace the governor spring if it is overstretched, and replace the linkages if they appear worn. **See Figure 7-21.**

INSPECTING GOVERNORS

① Secure spark plug lead away from spark plug

② Pull gently on throttle lever

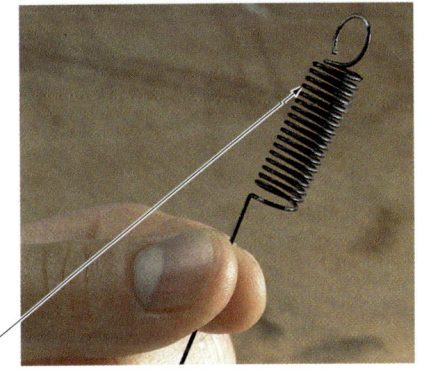

③ Replace governor spring if overstretched (shown)

Figure 7-21. A mechanical governor should be inspected for properly operating springs and linkages before maintenance service is performed.

... Project: Servicing Governor Systems ...

Troubleshooting Hunting and Surging Conditions

Erratic engine behavior is referred to as hunting and surging. For example, an engine may race or slow intermittently even when the load and the speed control settings are unchanged. Some engines have a separate governed idle spring and governed idle adjusting screw to prevent stalls under light loads.

To determine whether the source of hunting and surging (erratic engine behavior) is the carburetor or the mechanical governor, apply the following procedure:

1. Check that springs and linkages move freely and the governor spring is inserted properly on the governor lever arm.
2. Perform a static governor adjustment test by running the engine at each speed setting to determine when the hunting and surging condition occurs. With the engine running at idle, hold the throttle shaft against the idle speed screw to force the engine to idle. If a problem is present at "true idle" (when the throttle lever is against the idle speed screw or stop), the air-fuel mixture is likely the cause. An air leak or debris in the carburetor is probably causing the air-fuel mixture to fluctuate.
3. Use a screwdriver to remove and clean the carburetor.
4. If hunting and surging occurs at top no-load speed, move the throttle control to the high (rabbit) position. Hold the throttle shaft at the top no-load speed and observe the engine response. If hunting and surging continues, proceed to Step 5. If hunting and surging is eliminated, proceed to Step 6.
5. Clean and adjust the carburetor.
6. Lubricate the governor linkage with lubricating oil to eliminate any resistance and binding, or replace the governor spring(s) and retest.

Adjusting Static Settings on Mechanical Governors

A governor crank is the arm that protrudes from a crankcase. Servicing governor systems involves eliminating play in a mechanical governor between the governor crank and governor system components inside the crankcase.

To eliminate play in a mechanical governor between the governor crank and governor system components inside the crankcase, apply the following procedure (procedure does not apply if the engine has a pneumatic governor):

1. Use a socket wrench to loosen the governor arm clamp bolt on the governor crank until the governor lever moves freely.
2. Move the throttle plate linkage until the throttle plate is wide open. (To find the wide-open position, first position the throttle lever against the idle speed screw or a fixed stop plate. The throttle is wide open when it is all the way in the opposite direction.) Observe the direction of rotation of the governor arm as the throttle plate is moved to the wide-open position.

Chapter 7—Intermediate Small Engine Maintenance and Repair Projects 133

3. With the throttle plate wide open, use a nut driver or socket wrench to rotate the governor shaft in the same direction of governor arm travel.
4. Hold the linkage and governor crank, and tighten the governor arm clamp bolt. Manually move the linkage to ensure that there is no binding. *See Figure 7-22.*

ADJUSTING STATIC SETTINGS ON MECHANICAL GOVERNORS

① Loosen governor arm clamp bolt on governor crank

② Move throttle plate linkage until throttle plate is wide open

③ Use nut driver to rotate governor shaft in same direction of governor arm travel

④ Tighten governor arm clamp bolt

Figure 7-22. The static setting on a mechanical governor is adjusted to eliminate play between the governor crank (the arm that protrudes from the crankcase) and governor system components inside the crankcase.

...Project: Servicing Governor Systems...

TECH FACT

Many types of OPE have what is known as a "dead man lever," which is a speed control device that must be depressed to operate an engine. It automatically returns engine speed to idle or stops a moving piece of equipment, such as a blade, when it is no longer depressed.

Adjusting Governed Idle

Some engines contain a shorter, smaller secondary governor spring to discourage stalls when the engine is operating at idle under a light load. Under this condition, the secondary spring keeps the engine at a governed idle speed slightly above its true idle speed. The idle speed screw is always set at less than the governed idle speed of the engine. The procedure for adjusting governed idle varies depending on the engine model. The owner's/operator's manual should be consulted for the proper adjustment procedure for a particular engine. *Note:* The secondary spring affects all governor settings. If the governor on the engine has a secondary spring, adjust the governed idle before setting the top no-load speed.

Tang Bending and Other Adjustment Methods

A tang bending tool is the most common tool used for governor spring adjustment. It is a metal lever with forked ends used to grasp and bend the tabs, or tangs, on governor levers, spring anchors, and other engine parts. Bending a tang increases or decreases the extension of governor springs. If the governor lever has multiple spring holes, top no-load speed can be increased by selecting a hole that is farther from the pivot point on the governor lever. On some engines, an adjustment screw alters governor spring tension, increasing or decreasing top no-load speed. ***See Figure 7-23.*** However, a tang bending tool may still be required to make fine adjustments.

Chapter 7—Intermediate Small Engine Maintenance and Repair Projects   135

GOVERNOR ADJUSTMENT MECHANISM TYPES

Multiple governor springholes

Governor lever

SPRING HOLES ON GOVERNOR LEVER

Governor spring tension adjustment screw

Governor spring

ADJUSTMENT SCREW

Figure 7-23. Governor adjustment mechanisms on small engines can consist of either multiple spring holes on the governor lever or a spring tension adjustment screw.

... Project: Servicing Governor Systems

Setting Single-Spring Top No-Load Speed

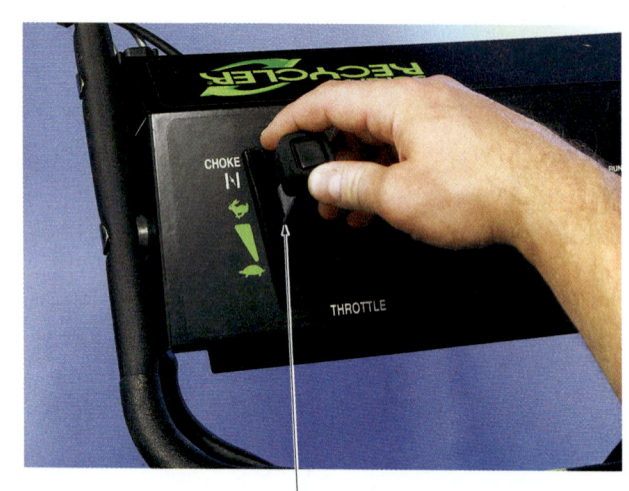

If an engine races when controls are set to a high position, the top speed under no-load conditions must be reduced. An authorized Briggs & Stratton service dealer must be consulted for the proper top no-load speed setting for specific engine models.

To set top no-load speed on a mechanical governor with one spring, apply the following procedure:

1. Attach a tachometer to the engine and run the engine for 5 min to reach its operating temperature.
2. Place the equipment on a hard level surface with the engine running and the controls set to high (rabbit icon).
3. Use a tang bending tool to decrease the top no-load speed by bending the tang toward the governor spring until the manufacturer-specified speed setting is attained. Increase the top no-load speed by lengthening the spring. *See Figure 7-24.*

SETTING SINGLE-SPRING TOP NO-LOAD SPEEDS

① Attach tachometer to engine and run engine for 5 min

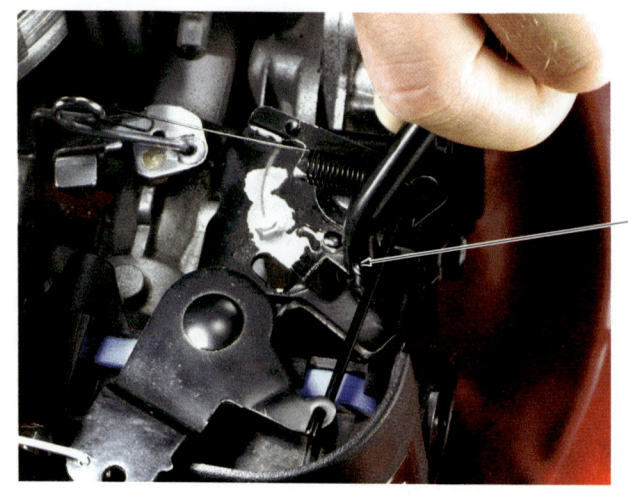

② Place equipment on hard level surface and set controls to high

③ Decrease top no-load speed by bending tang toward governor spring

Figure 7-24. If an engine races when controls are set to high, the top speed of the engine under no-load conditions must be reduced.

Setting Dual-Spring Top No-Load Speed

For a governor that contains two springs, the smaller, shorter spring is the secondary spring. This spring must be adjusted to prevent stalls.

To set top no-load speed on a mechanical governor with two springs, apply the following procedure:

1. Attach a tachometer to the engine.
2. With the engine running, bend the secondary spring tang so that there is no tension on the secondary spring.
3. Use a tang bending tool to bend the primary spring tang until the engine speed is 200 rpm below the manufacturer-specified top no-load speed.
4. Bend the secondary spring tang until the engine reaches its top no-load speed. Consult an authorized Briggs & Stratton dealer for the top no-load speed setting for the specific engine model. **See Figure 7-25.**

SETTING DUAL-SPRING TOP NO-LOAD SPEEDS

① Attach tachometer to engine

② Bend secondary spring tang (Secondary spring)

③ Bend primary spring tang

④ Bend secondary spring tang until engine reaches top no-load speed

Figure 7-25. When a governor contains two springs, the secondary spring must be adjusted to prevent stalls.

Project: Servicing Rewind Starters

- **Competency level:** intermediate
- **Tools needed:** socket wrench set, power drill, and 3/16″ drill bit
- **Estimated completion time:** 45 min

A rewind starting system, also known as a recoil system, is operated manually by pulling a rope. The rope is attached to a pulley and return spring that rotate the flywheel to start the spark plug firing and the engine running. A strong tug on the rope is usually required, since the flywheel must rotate fast enough to generate the high voltage necessary for ignition. If the rewind binds when it is pulled or does not rewind freely, it may need to be replaced. The entire rewind assembly should be replaced if any of the following conditions exist:

- rewind system is loud, binds, or feels rough when pulling rope
- crankshaft does not turn
- rope is frayed and worn along its entire length
- rope fails to rewind

On some engines, the rewind is spot-welded or riveted to the top of the engine shroud. On others, it is attached with nuts or bolts.

Replacing Rewind Starters

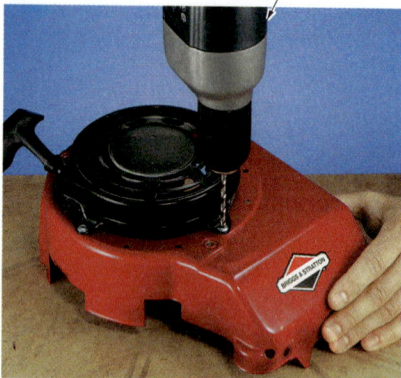

To replace a rewind starter, apply the following procedure:

1. Use a socket wrench to loosen the appropriate bolts, and remove the blower housing.
2. Use a socket wrench to remove the nuts or bolts on the rewind (if necessary), or use a power drill with a 3/16″ drill bit to drill out the rivets or spot welds, drilling only far enough to loosen them.
3. Install a replacement rewind from the OEM by inserting the mounting bolts from inside the blower housing. The bottom of the bolts should protrude out through the top of the shroud.
4. Place the replacement rewind over the bolts, and securely fasten a washer and nut on each bolt. ***See Figure 7-26.***

CAUTION: It is recommended that a small engine technician disassemble a rewind. A rewind assembly contains a pulley and spring that retract the rope after each pull. Replacing a rewind assembly requires special care and safety precautions must be taken because of the risk of serious injury from a spring or other flying parts.

REPLACING REWIND STARTERS

① Remove blower housing
② Drill out rivets or spot welds
③ Install replacement rewind
④ Fasten rewind assembly with washers, nuts, and bolts

Figure 7-26. The rewind starter must be replaced if the rope binds when it is pulled or does not rewind freely.

Chapter 7 Quick Quiz®

Advanced Small Engine Maintenance and Repair Projects

8

INTRODUCTION

In order for OPE to attain maximum life, a small engine must be properly maintained and repaired when damage occurs. Advanced small engine maintenance and repair projects typically require more effort and skill than basic and intermediate projects. Many advanced maintenance and repair projects for small engines can be performed in 60 min with a few common and specialty small engine hand tools. Advanced maintenance and repair projects include replacing drive discs, removing carbon deposits, servicing the electrical system, replacing the ignition system, and servicing valves. Most advanced projects require the adjusting and testing of components once they have been repaired.

WARNING: When performing any advanced maintenance and repair project, always wear safety glasses or goggles for protection from flying debris or splashing fuel.

Project: Replacing Drive Discs

- **Competency level:** advanced
- **Tools needed:** socket or open-end wrenches
- **Estimated completion time:** 30 min

A drive disc is a power transmission device that uses a friction disc to make contact between the driving and driven components of equipment. A friction disc has a rubberized friction compound molded to the circumference of a drive disc. The friction surface is placed in contact with a drive disc, which is connected to the engine. *See Figure 8-1.* Spring tension allows the friction disc to be located at various distances from the center of the engine drive disc. This results in variable speed control based on the position of the friction disc to the drive disc. With a drive disc system after a period of normal use, it is common for there to be slippage in the locomotion of equipment. When this occurs, the friction disc must be replaced.

To replace a drive disc, apply the following procedure:

1. Place the equipment on a hard, level surface such as a concrete driveway or garage floor.
2. Remove the spark plug wire and place it away from the spark plug.
3. Place and lock the shift lever in the fast-ground-speed position using the shift detent.
4. Stand the equipment up on one side to expose the undercarriage and drive disc.
5. Secure the equipment to a stationary surface with a rope or chain to ensure its stability while performing work.
6. Locate the drive disc assembly under the engine.
7. Use a socket or open-end wrench to remove the bolts that attach the disc to the assembly.
8. Place and lock the shift lever into the neutral position using the shift detent.
9. Replace the disc.
10. Place and lock the shift lever into the fast-ground-speed position using the shift detent.
11. Use a socket or open-end wrench to tighten the bolts that attach the disc to the assembly.
12. With the shift lever in the high position, verify that the disc contact point is farthest away from the center of the driven disc.
13. Move the shift lever from the high position to the neutral position and lock using the shift detent.
14. Verify that the new disc is not in contact with the drive disc attached to the engine by rotating the disc by hand.
15. Place and lock the shift lever in the high position using the shift detent.
16. Place the equipment back into the normal operating position on a hard, level surface.
17. Reconnect the spark plug wire to the spark plug, and return the equipment to service.

DRIVE DISCS

Figure 8-1. A friction surface is placed in contact with a drive disc, which is connected to an engine.

SMALL ENGINE AND EQUIPMENT MAINTENANCE

Project: Removing Carbon Deposits . . .

- **Competency level:** advanced
- **Tools needed:** parts cleaning solvent, socket wrench set, nylon-face hammer, putty knife, wooden or plastic blade scraper, wire brush or steel wool, brass wire brush, and torque wrench
- **Estimated completion time:** 60 min

A common by-product of combustion is carbon, which is the dark-colored soot that can collect and harden on the cylinder head, cylinder wall, piston, and the valves of an engine. Carbon is an undesirable element that is deposited in the combustion chamber of nearly every internal combustion engine. Although carbon is the main substance of the deposit, there are many other substances found in combustion deposits. Carbon deposits in the combustion chamber can affect engine performance, resulting in higher oil consumption, engine knocking, or engine overheating. The use of unleaded gasoline reduces carbon deposits. However, for the best operation, carbon deposits should be removed every 100 hr of engine operation by removing the cylinder head.

To remove engine components from an L-head engine for the removal of carbon deposits, apply the following procedure:

1. Remove the spark plug lead from the spark plug and the spark plug from the cylinder head.
2. Remove the muffler, muffler guard, and any other components that block access to the cylinder.
3. The cylinder head bolts near the muffler and exhaust port may be longer than other bolts. To avoid confusion, prepare a template. Draw a rough outline of the cylinder head on cardboard, and punch holes for each bolt location. Remove the cylinder head bolts, and insert them in the corresponding holes. Remove each cylinder head bolt and store in its respective slot in the cardboard template until the cylinder head is ready to be reinstalled.
4. Remove the cylinder head and head gasket. If the cylinder head sticks, tap it gently on the side with a nylon-face hammer. This should loosen the cylinder head enough to gently lift it off the engine. **CAUTION:** Do not pry off the cylinder head. This can damage the surface of the engine block or the cylinder head.
5. Remove and discard the old head gasket. *See Figure 8-2.*

REMOVING ENGINE COMPONENTS

① Remove spark plug lead from spark plug

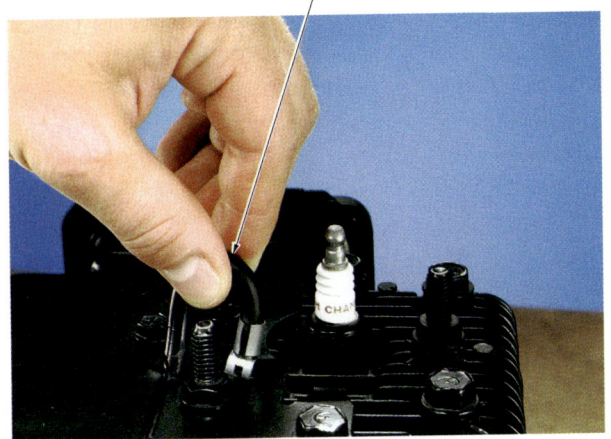

② Remove muffler and any other components blocking access to cylinder

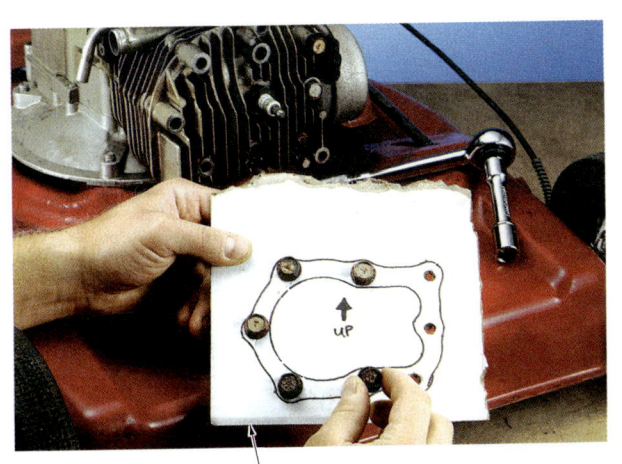

③ Sketch outline of cylinder head and bolt locations; insert bolts in hole outlines on sketch

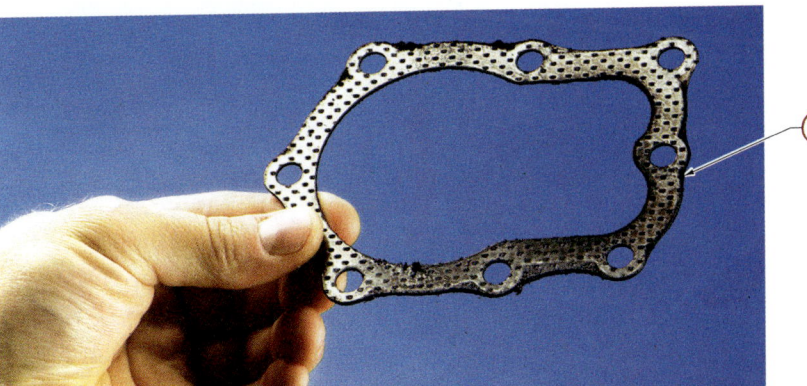

④ Remove cylinder head and head gasket

⑤ Discard old head gasket

Figure 8-2. Certain engine components must be removed before an engine can be cleaned of carbon deposits.

... Project: Removing Carbon Deposits ...

To remove carbon deposits, apply the following procedure:

1. Soak the cylinder head in a bath of ammonia-based household cleaner until the piston and engine-side combustion chamber are clean.

2. Rotate the flywheel until the piston is at top dead center (TDC) with both valves closed. Gently scrape any carbon deposits from the cylinder head using a wooden or plastic blade scraper. Take care not to dig the scraper into the aluminum. On stubborn deposits, use a putty knife, a wire brush, or steel wool, taking care not to bear down on the metal surfaces.

3. With the piston still at the top of the cylinder and the valves closed, use the same method to remove carbon deposits from the piston and the deck of the cylinder.

4. Rotate the crankshaft to open each valve, and carefully remove any visible carbon deposits on the valves and valve seats using a brass wire brush. **CAUTION:** Do not allow grit to fall into the valve chambers or between the piston and the cylinder wall.

5. Inspect the valves and valve seats to see if they are cracked, rough, or warped. Take damaged parts to an authorized Briggs & Stratton dealer for inspection before reassembling the head.

6. Using a scraper and/or solvent, remove any remaining carbon and residue left behind by the head gasket on the piston head, engine block, and cylinder wall. Clean the surfaces thoroughly before installing the new head gasket. Any debris or oil left on the cylinder head or engine block may prevent a tight seal and cause engine damage. *See Figure 8-3.*

CAUTION: Always wear protective eyewear and solvent-resistant gloves when removing carbon deposits. Consult an authorized Briggs & Stratton dealer for an all-purpose solvent that will not harm aluminum or plastic components or leave an unwanted residue.

Chapter 8—Advanced Small Engine Maintenance and Repair Projects 145

REMOVING CARBON DEPOSITS

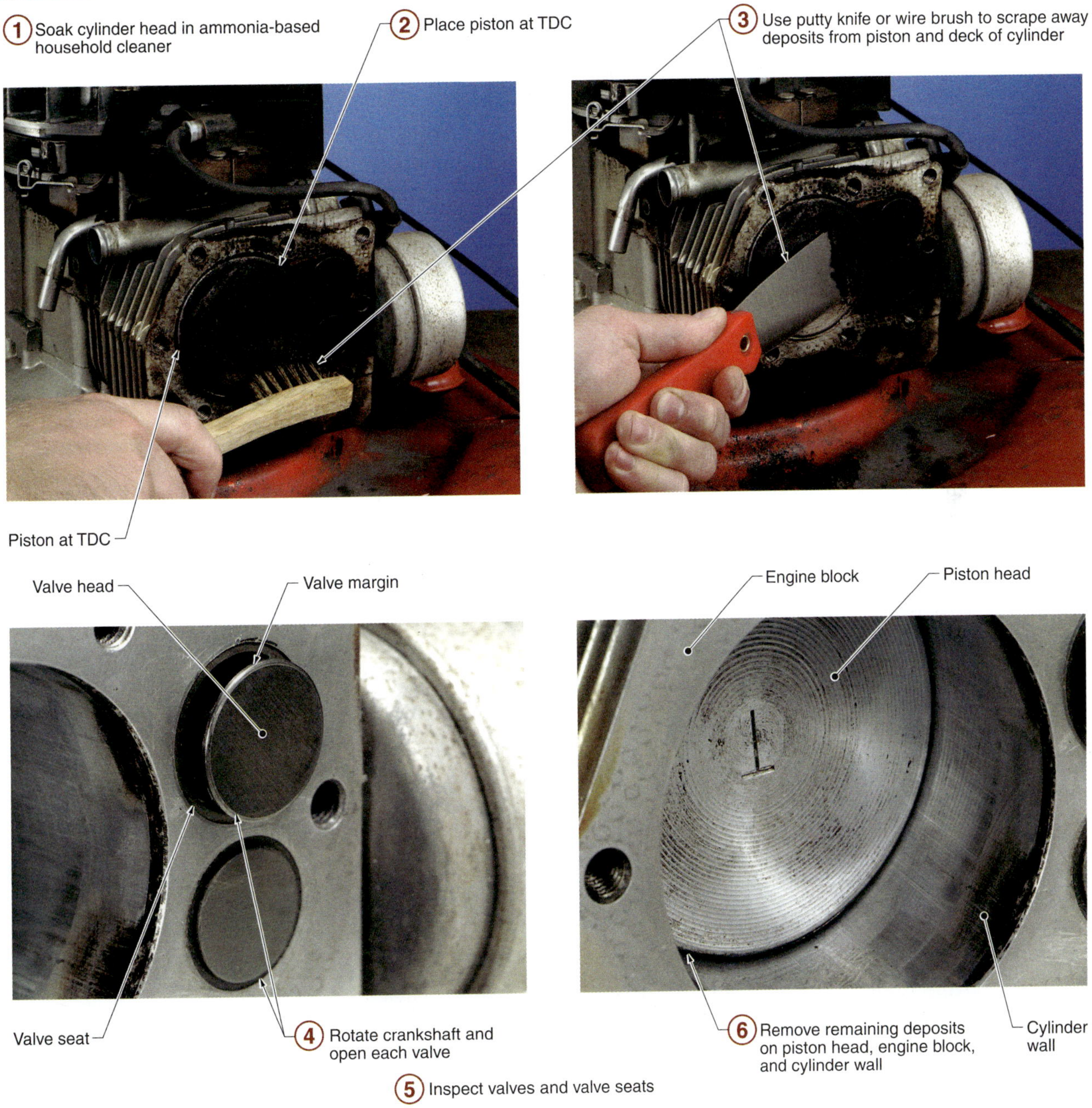

① Soak cylinder head in ammonia-based household cleaner
② Place piston at TDC
③ Use putty knife or wire brush to scrape away deposits from piston and deck of cylinder

Piston at TDC

Valve head — Valve margin
Engine block — Piston head

Valve seat
④ Rotate crankshaft and open each valve
⑤ Inspect valves and valve seats
⑥ Remove remaining deposits on piston head, engine block, and cylinder wall
Cylinder wall

Figure 8-3. For best operation, the cylinder head should be removed and cleaned of carbon deposits every 100 hours of operation.

146   SMALL ENGINE AND EQUIPMENT MAINTENANCE

. . . Project: Removing Carbon Deposits . . .

To reassemble a cylinder head, apply the following procedure:

1. Inspect the surfaces of the engine block deck, cylinder head, and new head gasket to ensure they are clean.
2. Position the new head gasket on the cylinder head. Do not use sealing compounds.
3. Set the cylinder head and head gasket on the engine deck, aligning the cylinder head with the gasket and the engine block.
4. Remove each head bolt from its slot in the cardboard template. Insert the bolts in their original locations, leaving them loose. Insert the other bolts in the same manner, making sure to attach any housings or brackets held in place by the head bolts.
5. Tighten the head bolts by hand. Do not use a wrench.
6. Tighten the cylinder head bolts in increments using a torque wrench. Turn each bolt a few turns, and then proceed to the next bolt until each bolt is just snug. For final tightening, use a torque wrench. Proceed to tighten in increments of roughly one-third the final torque. Consult the owner's/operator's manual for the final torque sequence and specifications. Avoid tightening a single bolt completely before tightening the other bolts. Uneven tightening is likely to warp the cylinder head.
7. Replace all components that were removed to gain access to the cylinder head and bolts.
8. Replace the spark plug and reconnect the spark plug lead. Place the unit back into service. *See Figure 8-4.*

Chapter 8—Advanced Small Engine Maintenance and Repair Projects

REASSEMBLING CYLINDER HEADS

① Inspect surfaces of engine block deck, cylinder head, and head gasket for cleanliness

② Position new head gasket on cylinder head

③ Align cylinder head with gasket and engine block

④ Remove bolts from cardboard template, and insert in original locations in cylinder head

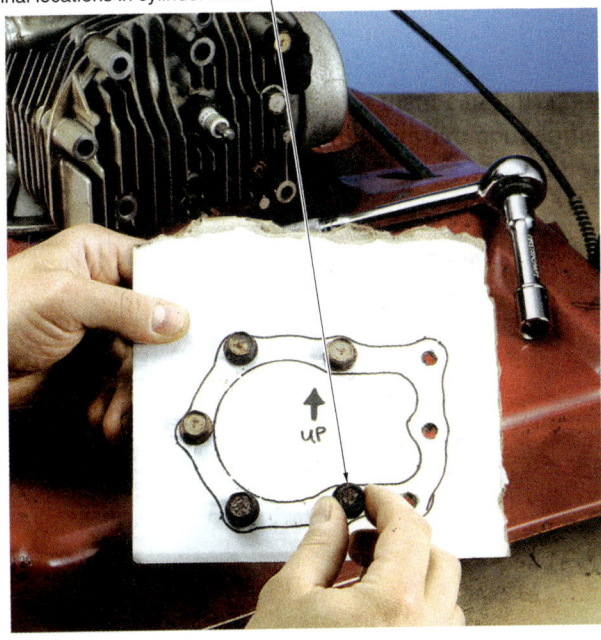

⑤ Hand-tighten head bolts

⑥ Tighten all head bolts completely with torque wrench

⑦ Replace all components removed to gain access to cylinder

⑧ Replace spark plug and spark plug lead

Figure 8-4. A cylinder head is easily reassembled once carbon deposits have been removed.

Project: Servicing Electrical Systems . . .

- **Competency level:** advanced
- **Tools needed:** digital multimeter (DMM), digital tachometer, shim, socket wrench set, and torque wrench
- **Estimated completion time:** 60 min

Small engines that are started with keys require electrical systems to charge the batteries and to power onboard electrical devices. When a sound, such as a groan or click, is made when starting a small engine equipped with an electric starter motor, there may be a problem with the electrical system. Electrical problems can prevent onboard electrical devices from operating. The proper electrical test can help identify the source of the problem. A reference chart can be used to review the most common testing connections for electrical systems on Briggs & Stratton engines and illustrate the proper method for using a digital multimeter (DMM) on an electrical system or component. *See Figure 8-5.*

A stator is an electrical component that has a continuous copper wire wound on separate metal stubs, which expose the winding to a magnetic field. On most engine models, the stator is mounted under the flywheel. The stator is not difficult to replace once the flywheel is removed. On some walk-behind lawn mowers the stator is mounted outside the flywheel. This location of the stator makes stator replacement even simpler.

Battery Location

Generally, an AC voltage test provides a value for the electrical power exiting the stator. Therefore, if the stator plugs into a regulator rectifier, an AC voltage test is first performed to verify the stator has the potential to provide current to the regulator. Next, the DC exiting the regulator/rectifier is tested. The DC eventually goes to the battery. The stator is replaced after it has been tested, and the air gap is adjusted, if necessary.

To determine the proper test for an alternator when servicing the electrical system of a small engine, apply the following procedure:

1. With the engine off, locate the thin wires extending from beneath the blower housing. These wires attach to the stator under the flywheel and deliver electrical current from the stator to the battery and other electrical devices.
2. Gently scrape away any engine paint as necessary to identify the true wire color. Note the color of the wires as well as the color of the wire connector, which is typically located an inch or two from the blower housing.
3. With Briggs & Stratton engines, locate the same wire-connector combination on the reference chart.

Note: The reference chart indicates the type of test to perform, how to set the DMM leads, the correct engine test speed, and the DMM readings. If the wiring for a specific engine is not on the chart, consult an authorized Briggs & Stratton dealer for the best test method.

Battery Safety

Small engines typically use lead-acid batteries, which store electrical energy using lead plates and sulfuric acid. The electrolyte fluid in the battery loses its sulfuric acid and gains water as the battery is discharged. Battery electrolyte is extremely corrosive and can severely burn the eyes and skin. Batteries also produce hydrogen gas that can cause an explosion if ignited by a spark or open flame. The hazards associated with batteries can be minimized by observing the following precautions:

- Follow all manufacturer recommended procedures for battery charging, installation, removal, and disposal.
- Always hold a battery upright to avoid spilling electrolyte.
- Wear protective eyewear and clothing when handling batteries.
- If electrolyte spills on skin or splashes in eyes, flush immediately with cold water and contact a physician.
- Perform battery service only in a well-ventilated area, away from sources of sparks or flames.

ELECTRICAL SYSTEM TESTING REFERENCE CHARTS

AC VOLTAGE TEST

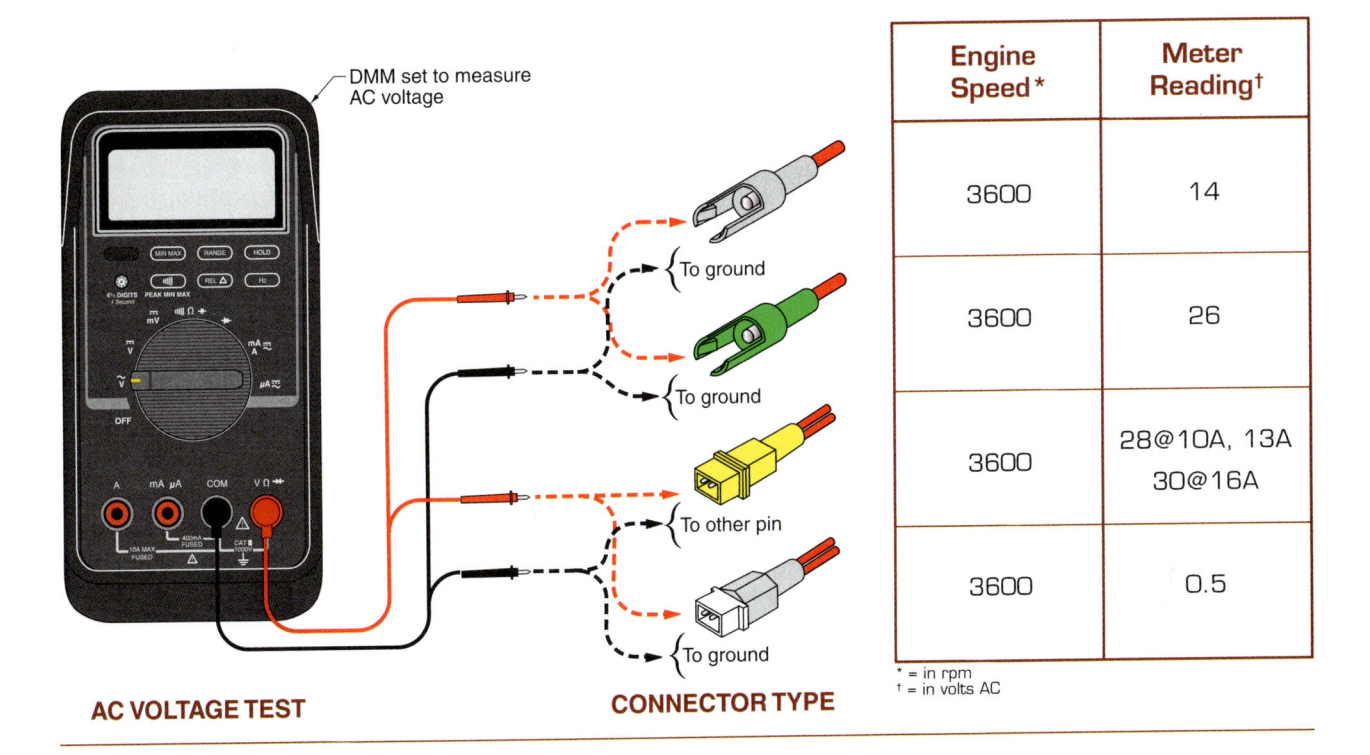

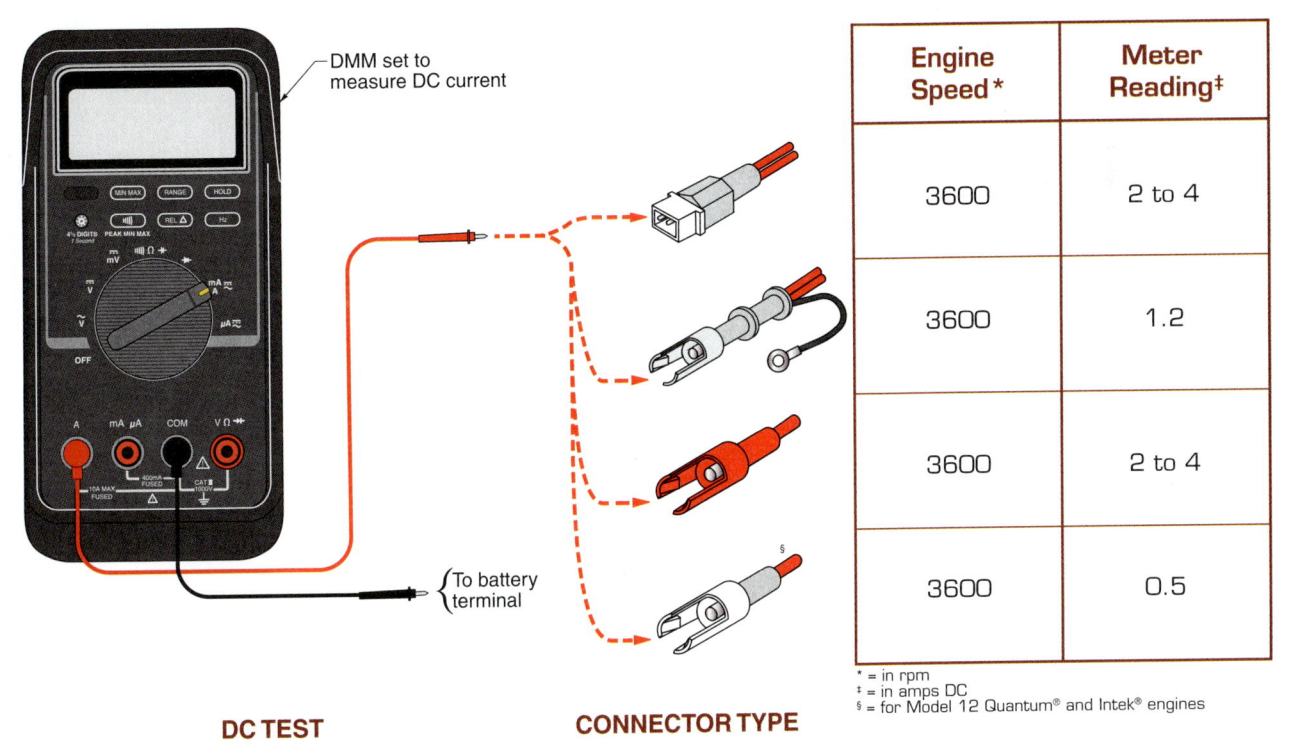

CONNECTOR TYPE

Engine Speed*	Meter Reading†
3600	14
3600	26
3600	28@10A, 13A 30@16A
3600	0.5

* = in rpm
† = in volts AC

DC TEST

CONNECTOR TYPE

Engine Speed*	Meter Reading‡
3600	2 to 4
3600	1.2
3600	2 to 4
3600	0.5

* = in rpm
‡ = in amps DC
§ = for Model 12 Quantum® and Intek® engines

Figure 8-5. A reference chart can be consulted to review and illustrate the proper method for using a DMM on a Briggs & Stratton engine electrical system or component.

...Project: Servicing Electrical Systems...

To test AC voltage, apply the following procedure:

1. Set the DMM selector switch to measure AC voltage. Insert the black meter lead into the DMM receptacle labeled "COM," and hold the probe tip against a good engine ground such as an engine mounting bolt or cylinder fin.
2. Insert the red meter lead on the DMM into the AC voltage receptacle (V/Ω/→) of the tester, and hold the probe tip against the appropriate stator output wire.
3. Start the engine and set the speed control to the fast position. Check the engine rpm using a digital tachometer to verify the speed setting. Note the output on the DMM display, and replace the stator if the reading is incorrect.
4. Shut off the engine and disconnect the DMM from the equipment. *See Figure 8-6.*

ALTERNATING CURRENT (AC) VOLTAGE TESTS

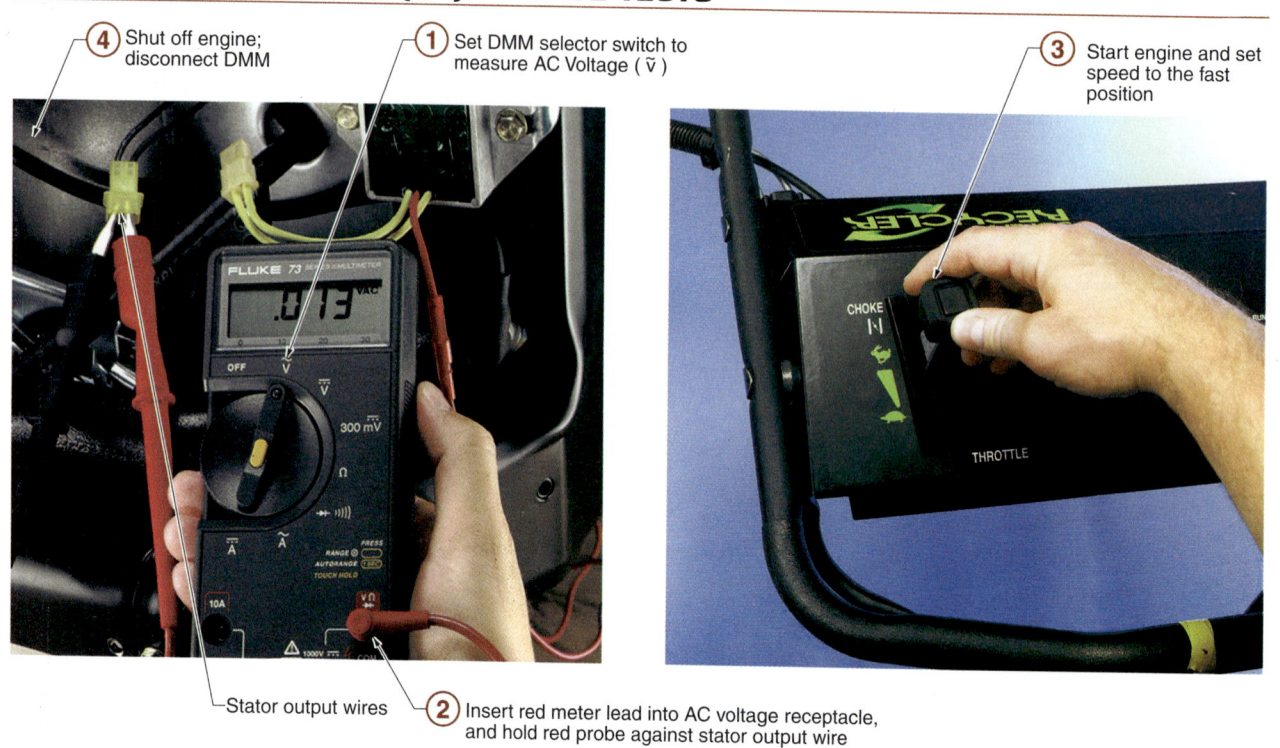

Figure 8-6. An AC voltage test is first performed to verify that a stator has the potential to provide current to a regulator.

Chapter 8—Advanced Small Engine Maintenance and Repair Projects 151

To test DC, apply the following procedure:

1. Set the DMM selector switch to measure DC. Insert the black meter lead into the DMM receptacle labeled "COM," and hold the probe tip against the positive (+) terminal of the battery. *Note:* The battery must be grounded to the equipment frame or the engine block to create a complete circuit.
2. Insert the red meter lead into the current (A) receptacle of the DMM, and hold the probe tip against the appropriate stator output wire.
3. Start the engine and set the speed control to the fast position. Check the engine rpm using a digital tachometer to verify the speed setting. Note the output on the DMM display, and replace the stator if the reading is incorrect.
4. Shut off the engine and disconnect the DMM from the equipment. *See Figure 8-7.*

TECH FACT
It is not necessary to perform AC voltage tests on stators that have diodes in line with output leads because the stators are already rectified by the diodes.

Note: When taking current measurements higher than 1A, a DMM with a current clamp attachment or clamp meter must be used.

DIRECT CURRENT (DC) TESTS

① Set DMM selector switch to measure DC ($\widetilde{A}$); hold black probe against positive (+) battery terminal

② Insert red meter lead into current (A) receptacle; hold probe tip against stator output wire

③ Start engine and set speed control to fast position

④ Shut off engine and disconnect DMM

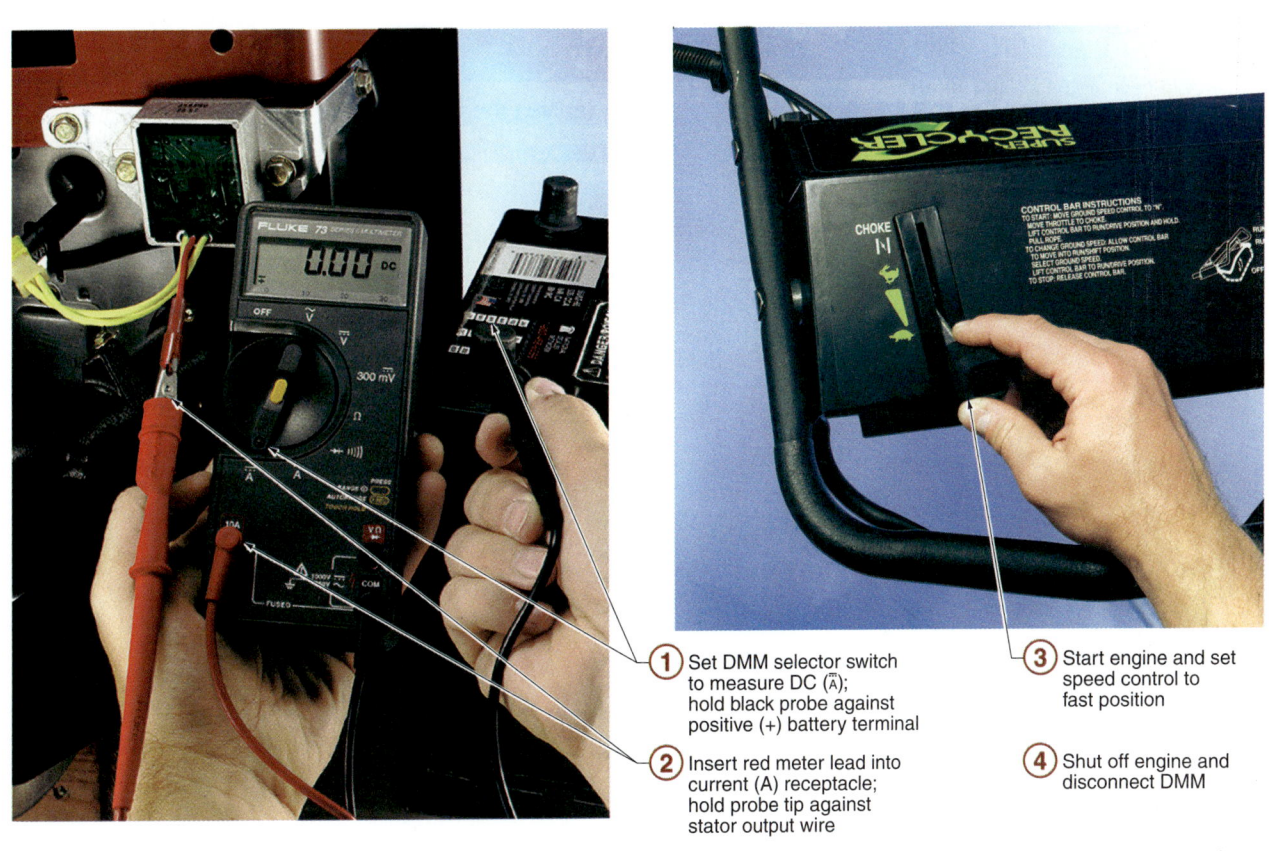

Figure 8-7. DC exiting a regulator/rectifier is tested to verify that the electrical system of an engine is operating properly.

...Project: Servicing Electrical Systems...

To replace a stator located under a flywheel, apply the following procedure:

1. With the flywheel, screen, and blower housing removed, note the path of the stator wires (typically under one coil spool and between the starter and starter drive housing). *Note:* With most OPE, the blower housing, rotating screen, rewind clutch, and flywheel must be removed to access the stator.

2. Remove the ground wire or rectifier assembly (if equipped) from the starter drive housing. Remove the stator mounting screws and bushings.

3. Before installing a new stator, locate the stator wires against the cylinder, and ensure the wires remain clear of the flywheel.

4. Install a new stator assembly, ensuring the output wires are properly positioned. While tightening the mounting screws, push the stator toward the crankshaft to take up clearance in the bushing. Tighten the screws to 20 in.-lb.

5. Reinstall the flywheel, screen, and blower housing. Attach the ground wire or rectifier assembly (if equipped) to the drive housing. *See Figure 8-8.*

REPLACING STATORS LOCATED UNDER FLYWHEELS

① Remove flywheel and any other components blocking stator access

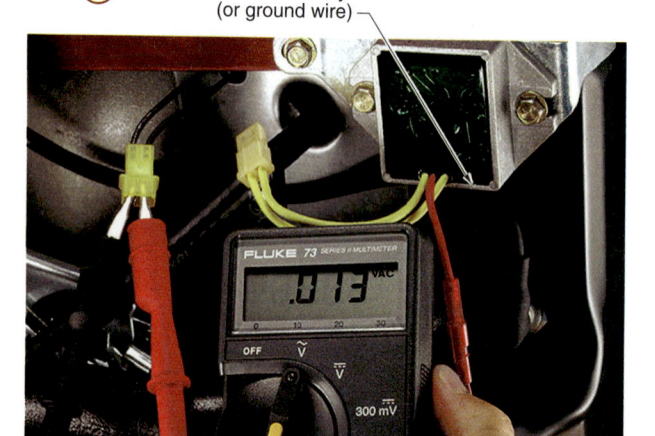

② Remove rectifier assembly (or ground wire)

③ Locate stator wires and keep them clear of flywheel

④ Install new stator assembly

⑤ Replace flywheel, screen, and blower housing

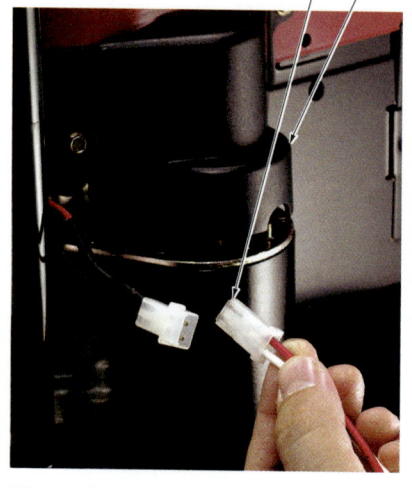

Figure 8-8. A stator located under a flywheel is accessed by removing the flywheel, screen, and blower housing.

To replace a stator mounted outside a flywheel (an external stator), apply the following procedure:

REPLACING EXTERNAL STATORS

1. Disconnect the stator output wire from the wires leading to the battery or other electrical devices.
2. Rotate the flywheel until the magnets are positioned away from the stator. Loosen the stator mounting bolts, and remove the stator from the engine.
3. With the flywheel magnets positioned away from the stator, install the new stator, leaving a wide gap between the stator and the flywheel. Tighten one of the mounting bolts.
4. Reattach the stator output wires. **See Figure 8-9.**

Figure 8-9. When a stator is mounted outside a flywheel, the stator is easily removed by loosening the two mounting bolts and lifting it off the engine.

...Project: Servicing Electrical Systems...

To adjust the air gap on an external stator, apply the following procedure:

1. Rotate the flywheel until the magnets are positioned away from the stator.
2. Loosen both stator mounting bolts, and move the stator away from the flywheel. Tighten one of the mounting bolts.
3. Place a shim or piece of cardboard of the proper air gap thickness between the stator and the flywheel. *Note:* The air gap between the stator and the flywheel must be set precisely for the stator to function properly. Many stators require a 0.010″ air gap. Consult an authorized Briggs & Stratton dealer for the proper gap for a specific stator.
4. Rotate the flywheel until the magnets are adjacent to the stator.
5. Loosen the tightened bolt, and allow the magnets to pull the stator until it is flush with the shim.
6. Use a torque wrench to tighten both mounting bolts to 25 in.-lb.
7. Rotate the flywheel while pulling on the shim to release it. *See Figure 8-10.*

Electrical System Service

ADJUSTING AIR GAPS ON EXTERNAL STATORS

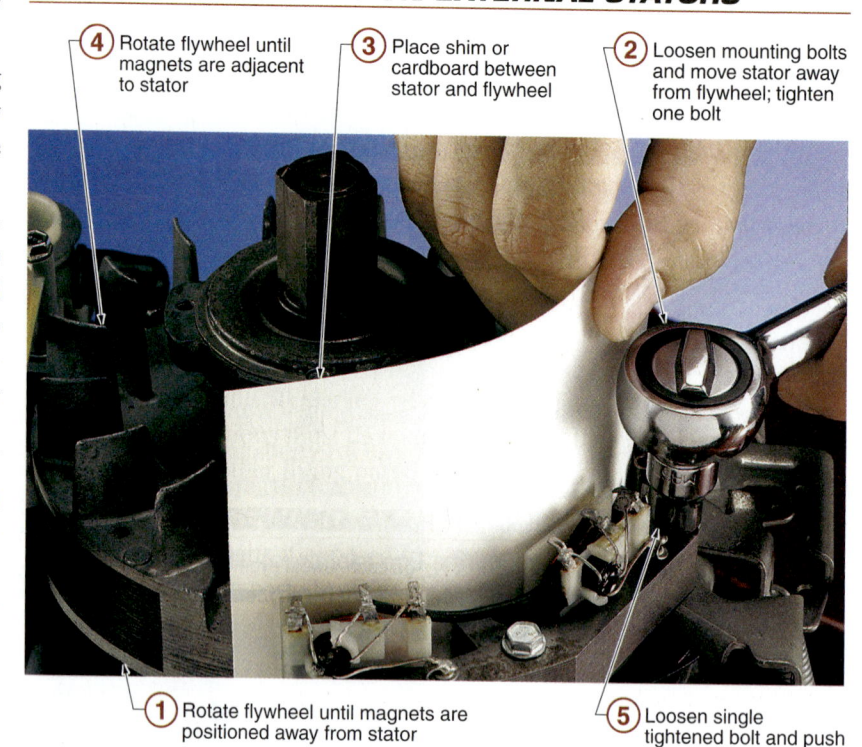

① Rotate flywheel until magnets are positioned away from stator
② Loosen mounting bolts and move stator away from flywheel; tighten one bolt
③ Place shim or cardboard between stator and flywheel
④ Rotate flywheel until magnets are adjacent to stator
⑤ Loosen single tightened bolt and push stator flush with shim
⑥ Tighten bolts with torque wrench
⑦ Rotate flywheel and remove shim

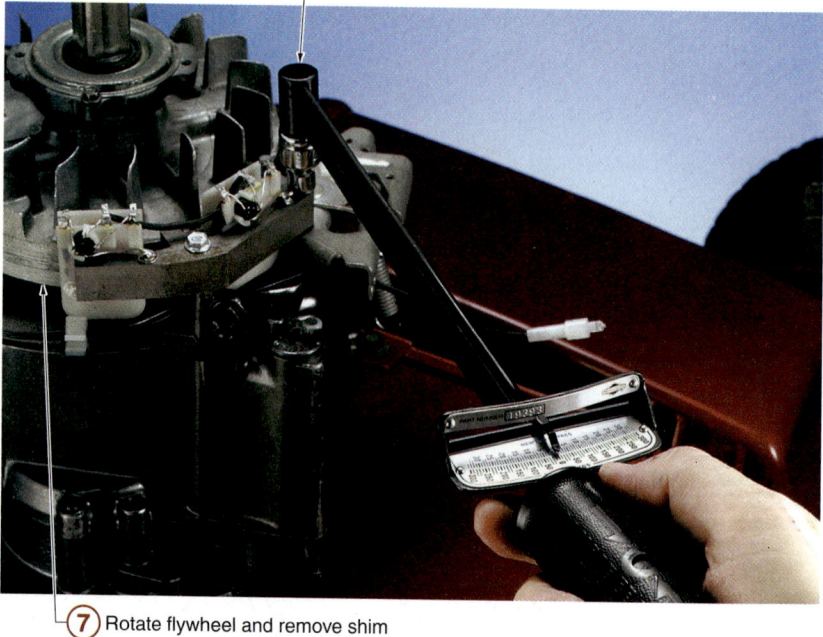

Figure 8-10. The air gap between a stator and flywheel must be set precisely for the stator to function properly.

Project: Replacing Ignition Systems . . .

- **Competency level:** advanced
- **Tools needed:** bench vise, ³⁄₁₆″ pin punch, razor blade or utility knife, feeler gauges, socket wrench, silicone sealer, soldering iron, 60/40 solder, spark plug tester, and wire cutters
- **Estimated completion time:** 60 min

Modern small engines contain a solid-state ignition armature mounted adjacent to the flywheel. The only moving parts in the ignition system are the magnets mounted in the flywheel. The magnets interact with the armature to produce electrical current. Most ignition armatures are designed to be replaced if they fail, not repaired. Even for armatures that can be repaired, replacing a failed armature is typically easier than repairing the part.

Most engines built through the early 1980s contain a set of mechanical points, known as breaker points, under the flywheels. The points open and close an electrical circuit for ignition. The ignition reliability of a single-cylinder Briggs & Stratton engine equipped with breaker points and two-leg armatures can be improved by installing a solid-state ignition conversion kit to bypass the breaker points. Consult an authorized Briggs & Stratton dealer for the proper conversion kit. *See Figure 8-11.*

EVOLUTION OF IGNITION ARMATURES

BREAKER POINTS IGNITION ARMATURE

EARLY SOLID-STATE IGNITION ARMATURE — Replaceable ignition module

MODERN SOLID-STATE IGNITION ARMATURE — Composite ignition module

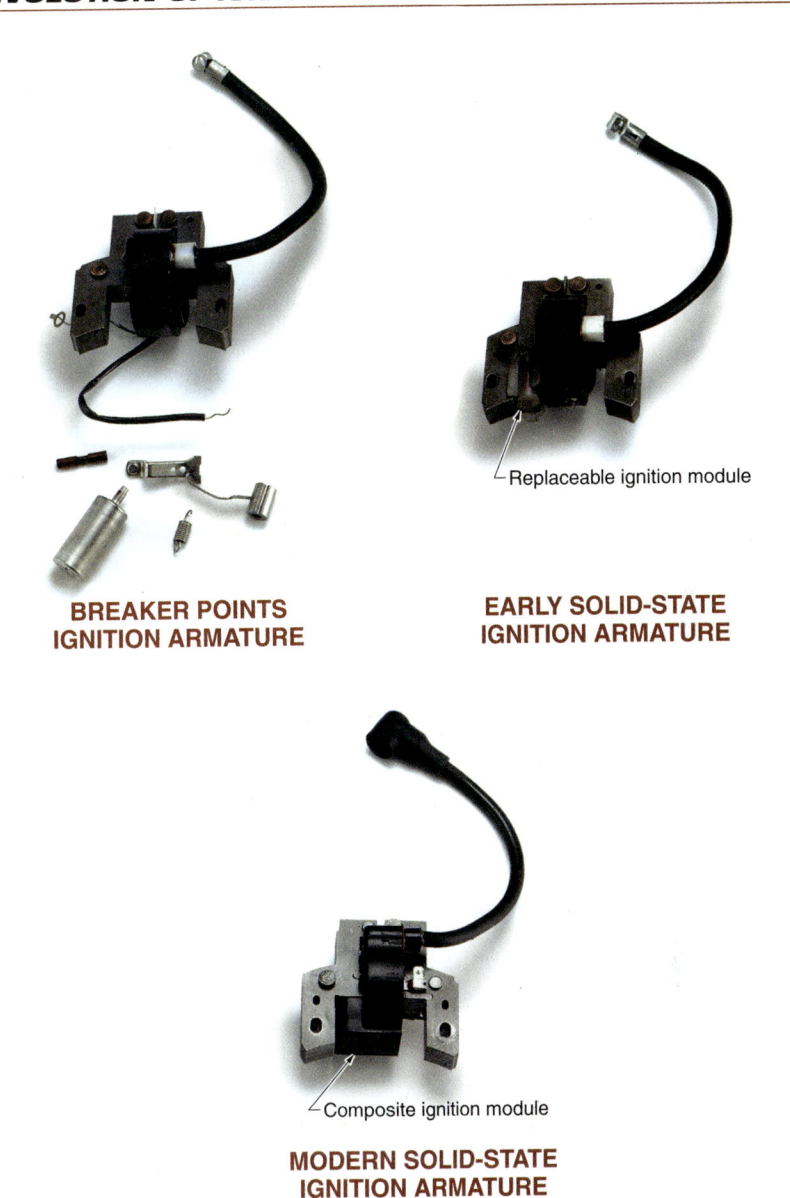

Figure 8-11. Ignition armatures have evolved from being of the breaker-points to solid-state types.

... Project: Replacing Ignition Systems ...

Ignition system service includes replacing ignition armatures, testing stop switches, and retrofitting old model ignition armatures. Before replacing a faulty ignition armature, the ignition system should be tested with a spark tester. Also, the ignition system should be tested for faulty electrical switches that could be the source of a problem.

To replace an ignition armature, apply the following procedure:

1. Remove the old ignition armature mounting bolts. Disconnect the stop switch wire from the flywheel brake and remove the armature. **See Figure 8-12.**
2. Attach a replacement armature from the original engine manufacturer (OEM) using mounting bolts. Push the armature away from the flywheel, and tighten one bolt.
3. Rotate the flywheel so that the magnets are on the opposite side of the ignition armature.
4. Place the appropriate size feeler gauge between the rim of the flywheel and the ignition armature. While holding the feeler gauge, rotate the flywheel until the magnets are directly adjacent to the armature.
5. Loosen the tightened bolts, and allow the magnets to pull the ignition armature against the flywheel and shim. Tighten both mounting bolts, and rotate the flywheel until the shim slips free.

Note: An ignition armature must be set to a precise gap from the flywheel. Consult an authorized Briggs & Stratton dealer for the proper gap for a specific engine. Armature gaps commonly range from 0.006″ to 0.014″. Ignition armatures are often packaged with shims to assist in setting the gaps.

Chapter 8—Advanced Small Engine Maintenance and Repair Projects 157

REPLACING IGNITION ARMATURE

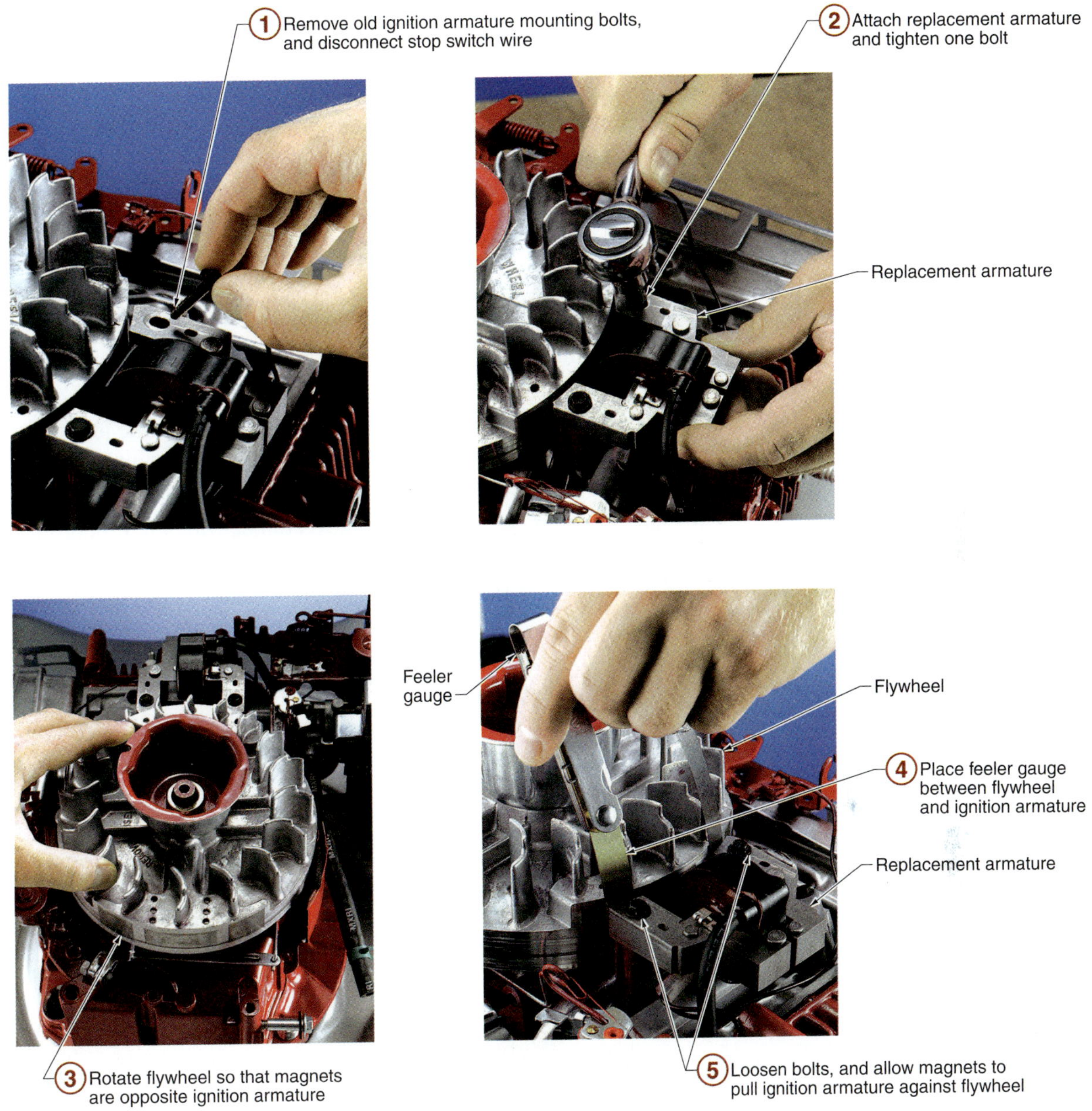

① Remove old ignition armature mounting bolts, and disconnect stop switch wire

② Attach replacement armature and tighten one bolt

— Replacement armature

— Feeler gauge

— Flywheel

④ Place feeler gauge between flywheel and ignition armature

— Replacement armature

③ Rotate flywheel so that magnets are opposite ignition armature

⑤ Loosen bolts, and allow magnets to pull ignition armature against flywheel

Figure 8-12. An ignition armature is installed by using a socket wrench and feeler gauges.

...Project: Replacing Ignition Systems...

To test a stop switch, apply the following procedure:

1. Insert the spark plug lead on one end of a spark tester and attach the alligator clip of the tester to a known ground such as an engine bolt.
2. Place the equipment stop switch control in the OFF (stop) position. If the engine is not connected to the equipment, ground the stop switch wire to the cylinder.
3. Attempt to start the engine using the rewind cord or key, if equipped. There should be no spark. If a spark appears, inspect the stop switch for damage. Consult an authorized Briggs & Stratton dealer if there is a faulty switch.
4. Place the stop switch control in the run or start position. If the engine is not connected to the equipment, make sure the stop switch wire is not grounded.
5. Attempt to start the engine. A spark should be visible in the tester. If no spark appears, check for broken wires, short circuits, grounds, or a defective stop switch.
6. Once the stop switch is working, reconnect the spark plug lead. **See Figure 8-13.**

TESTING STOP SWITCH

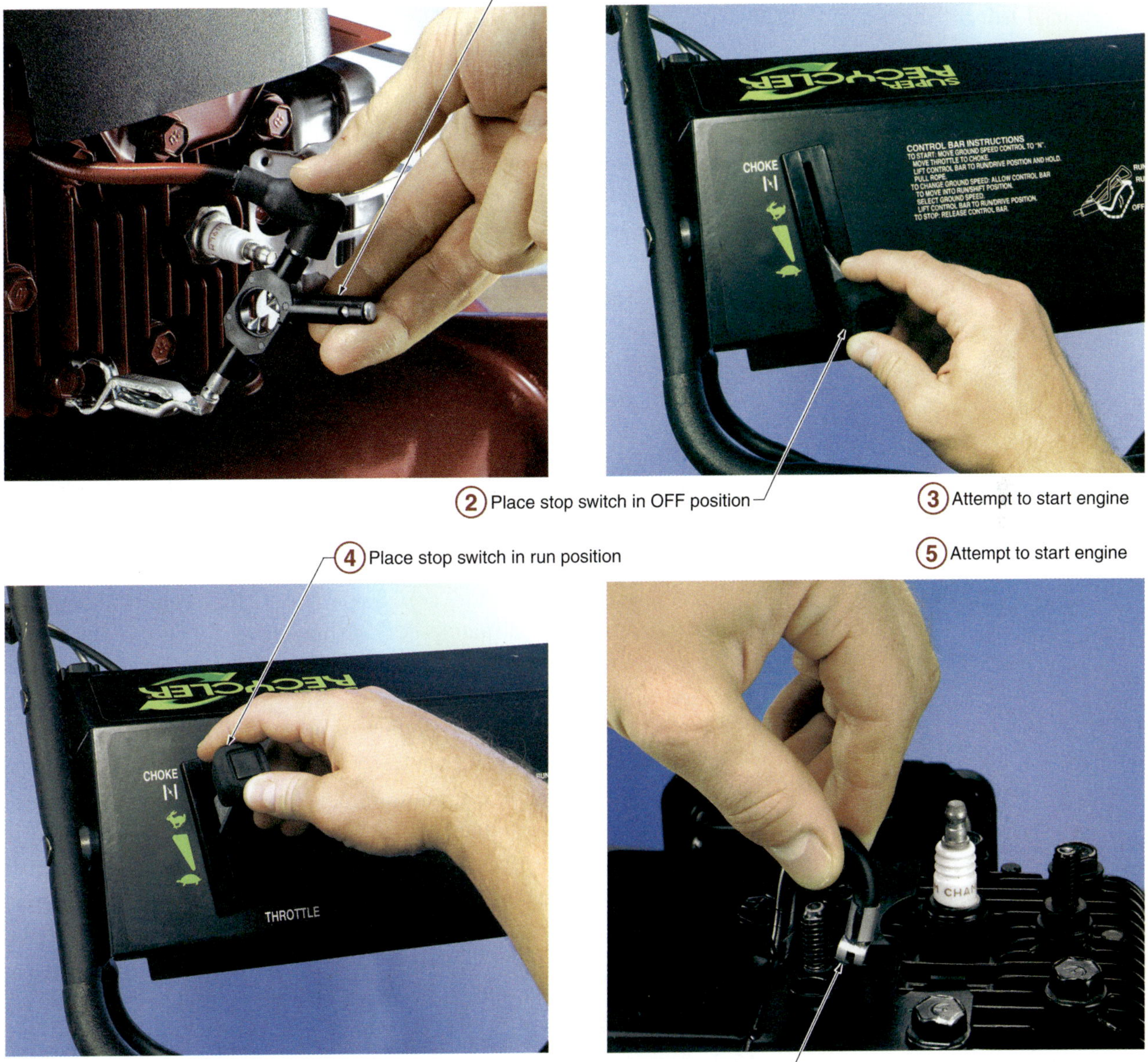

1. Disconnect spark plug lead from spark plug and attach to spark plug tester
2. Place stop switch in OFF position
3. Attempt to start engine
4. Place stop switch in run position
5. Attempt to start engine
6. Reconnect spark plug lead to spark plug

Figure 8-13. A stop switch is tested by attempting to start the engine while it is connected to a spark tester.

...Project: Replacing Ignition Systems...

To retrofit an old model ignition armature, apply the following procedure:

1. Disconnect the spark plug lead and secure it away from the plug. Remove the flywheel and discard the old flywheel key.
2. Cut the armature primary and stop switch wires as close as possible to the dust cover. Remove the dust cover, points, and plunger. Plug the plunger hole with the plug from the conversion kit.
3. Loosen the screws and remove the armature. Cut the primary wire of the armature to a 3″ length and strip ⅝″ off the outer insulation.
4. Use a utility knife or razor blade to remove the varnish insulation underneath the plastic insulation. Take care not to nick or cut the wire.
5. Install the conversion module, and modify the air vane brackets or guides for clearance as required.
6. Place a pin punch in the jaws of a bench vise and push open the spring-loaded wire retainer by pressing down on the punch. With the wire retainer slot open, insert the primary wire of the armature and a new stop switch wire (if required) together with the module primary wire. Release the wire retainer, locking the wires in place.
7. Secure the wires by soldering the ends together with 60/40 solder.
8. Twist the armature ground wire and module ground wire together (two turns) close to the armature coil. Solder the twisted section, taking care not to damage the armature coil casing. *Note:* Avoid crossing these wires with those inserted in the wire retainer in Step 6.
9. Remove the shortest ground wire by cutting it off close to the soldered connection. Attach the twisted ground wires to the armature coil using a generous amount of silicon sealer to protect against vibrations.
10. Use a screw to attach the armature/module ground wire to the armature. Fasten the armature to the engine so that the wire retainer is toward the cylinder.
11. Remove the remainder of the original stop switch wire as close as possible to the terminal on the engine. Route the new wire from the module, following the same path as the original.
12. Fasten the new wire in place. Verify that the wire does not interfere with the flywheel.
13. Install the flywheel using the replacement flywheel key from the kit. Torque the flywheel nut or rewind clutch. Set the armature air gap and test the stop switches. **See Figure 8-14.**

RETROFITTING OLD MODEL IGNITION MODULES

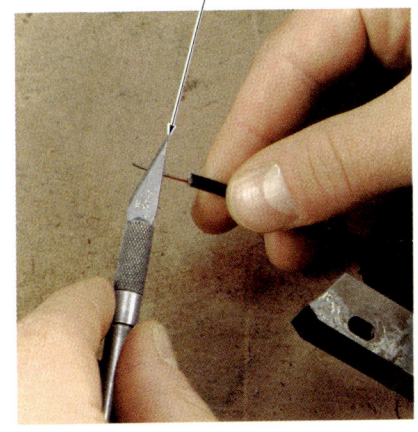

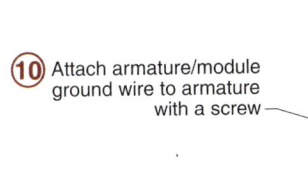

1. Remove flywheel and discard key
2. Cut armature primary stop switch wires
3. Cut armature primary wire to 3″ length
4. Remove varnish from stripped wire with razor knife
5. Install conversion module
6. Place armature primary wire, new stop switch wire, and module primary wire in wire retainer
7. Solder wire ends together
8. Twist armature ground wire and module ground wire together, and solder twisted section
9. Cement soldered wire to armature coil
10. Attach armature/module ground wire to armature with a screw
11. Remove remainder of original stop switch wire
12. Fasten new stop switch wire
13. Install flywheel and adjust armature air gap

Figure 8-14. To improve engine operation and extend the operating life of the engine, an old ignition module can be replaced with a more modern ignition module.

Project: Servicing Valves . . .

- **Competency level:** advanced
- **Tools needed:** valve spring compressor, feeler gauge set, hex wrench set, needle-nose pliers, nut driver set, calipers, safety eyewear, socket wrench set, torque wrench, lapping compound, and valve lapping tool
- **Estimated completion time:** 90 min

A four-stroke cylinder small engine contains one intake valve and one exhaust valve per cylinder. *See Figure 8-15.* Intake valves open to allow the air-fuel mixture to enter the combustion chamber. Exhaust valves open to allow spent fuel gases to exit the engine. The intake and exhaust valves close to seal the combustion chamber for the compression stroke of the piston. Valve springs push the valves toward the closed position so that they open only at precisely timed intervals. The valves are pushed open by tappets that ride on the lobes of the camshaft. The camshaft turns along with the crankshaft. The camshaft and crankshaft are driven by and synchronize with the piston.

A valve spring compressor is an essential tool for removing and installing valves on L-head engines and some overhead valve (OHV) engines. Valves control the flow of fuel vapor into a combustion chamber and the flow of exhaust gases leaving an engine. Faulty or dirty valves may stick and can develop pits, cracks, or grooves that cause the engine to lose power and fuel to lose efficiency. Valves should be carefully inspected once removed from the engine. If the valves are not badly worn and the parts are not damaged, the valves and seats can be tuned so that the valves seal effectively.

A valve has a stem, neck, head, and face. The valve stem moves in a valve guide that is machined directly in the cylinder block or in a replaceable bushing. Also, each valve moves through a valve spring adjacent to the guide that pulls the valve toward the closed position and holds the valve face against the valve seat. Each valve spring is held in place by a valve spring retainer. Some valve assemblies also include rotators, which are circular components that turn the valves slightly in each cycle to ensure even wear patterns on the valves and seats.

FOUR-STROKE CYLINDER-ENGINE VALVES

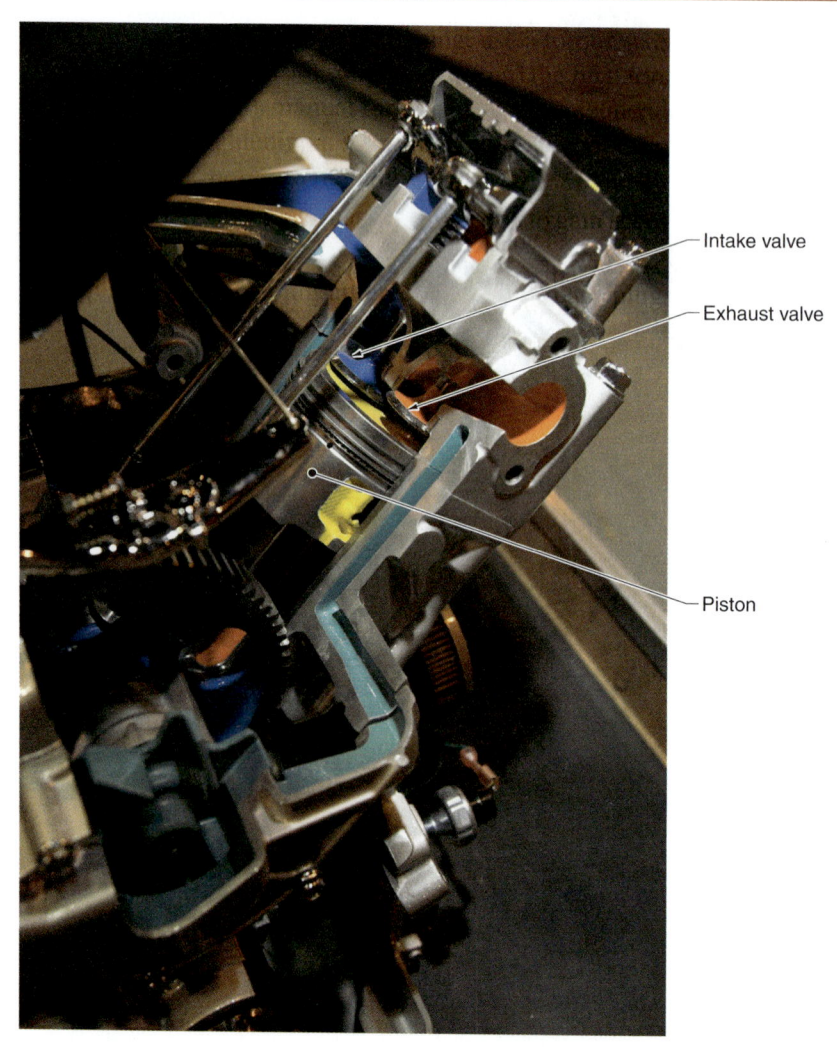

Figure 8-15. The valve design for a four-stroke cylinder small engine includes one intake valve and one exhaust valve per cylinder.

The valves of L-head engines are located on one side of the cylinder. The valve stems run through the cylinder block, parallel to the piston. The valves of overhead valve engines are located in the cylinder head. *See Figure 8-16.* Overhead valves are pushed open by pivoting rocker arms operated by push rods. The push rods, in turn, are pushed toward the rocker arms by tappets. The slightly more complex design yields greater power.

Valve lapping is done to repair minor scoring and pitting of the valve face and valve seat and to restore the ability of the valve to seal the combustion chamber. The valve lapping procedure involves gently rotating the valve in the seat with a layer of lapping compound, which is a fine but abrasive paste, between the valve and seat. A lapping tool is used to hold and rotate the valve. During lapping, progress must be monitored often. Otherwise, it is easy to remove not only the carbon buildup but also the metal, further damaging the valve or seat.

Since lapping removes a small amount of material from the surfaces of the valve face and valve seat, the tappet clearances (the spacing between the valve stem and the tappet) may need to be adjusted after lapping and reinstalling valves. *See Figure 8-17.* An authorized Briggs & Stratton dealer should be consulted for the correct tappet clearance for a specific engine.

VALVE DESIGN

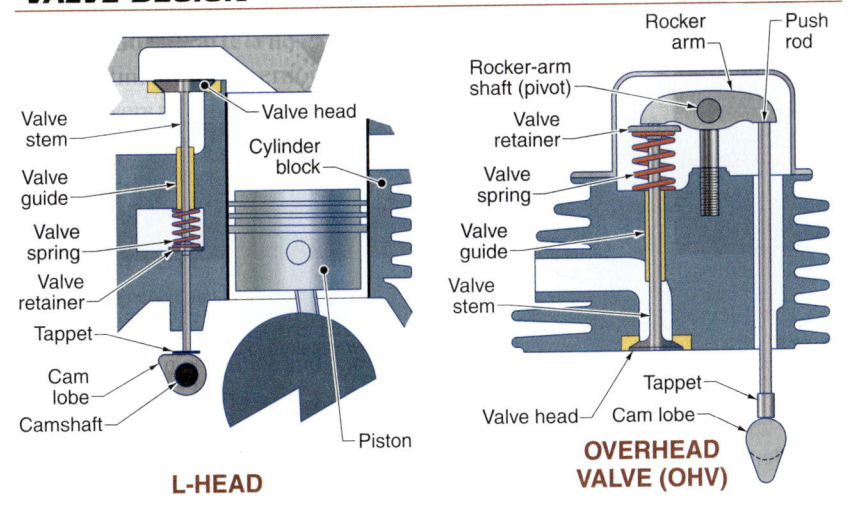

Figure 8-16. The valves of L-head engines are located on one side of the cylinder. The valves of overhead valve engines are located in the cylinder head.

TAPPET CLEARANCES

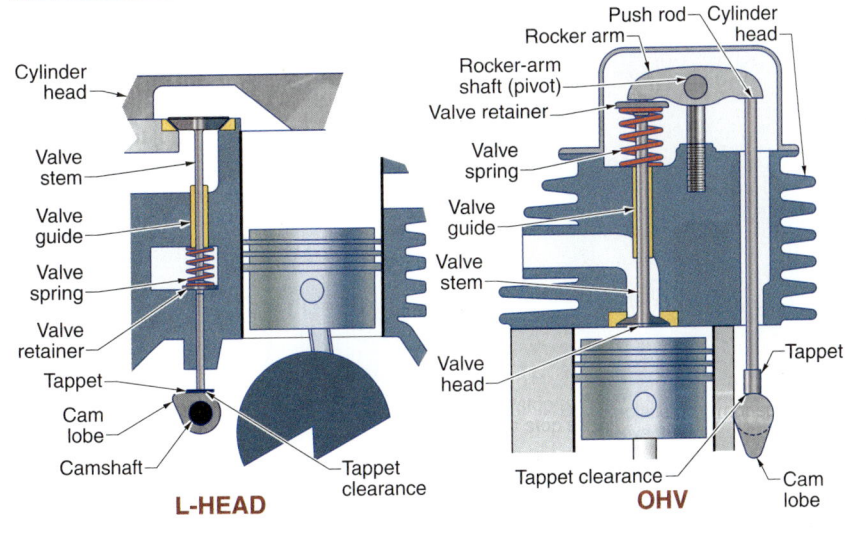

Figure 8-17. A tappet clearance is the spacing between the valve stem and the tappet. It may need to be adjusted after lapping and reinstalling a valve.

Valve Function

... Project: Servicing Valves ...

Machining Valves

Engine valve systems may function for years without bent, cracked, or damaged parts. However, they may have become worn and no longer form good seals when closed. The result is wasted fuel and loss of power each time the engine runs. An engine with old valves should be taken to an authorized Briggs & Stratton dealer for the restoration of valve seats or installation of new seats.

Servicing engine valves requires gaining access to the valve chamber; removing, inspecting, and lapping the valves; adjusting the tappet clearances; and reinstalling the valves. The valves may also require adjustment after being reinstalled.

To access a valve chamber, apply the following procedure:

1. Remove the muffler, crankcase breather, and any other components that block access to the valve chamber. ***See Figure 8-18.***
2. Remove the cylinder head bolts. Label the bolts, if necessary, to ensure proper installation later, since they may be of different lengths. *Note:* Always wear safety goggles or glasses when removing and installing valves.

ACCESSING VALVE CHAMBER

① Remove muffler and crankcase breather

② Remove cylinder head bolts and cylinder head

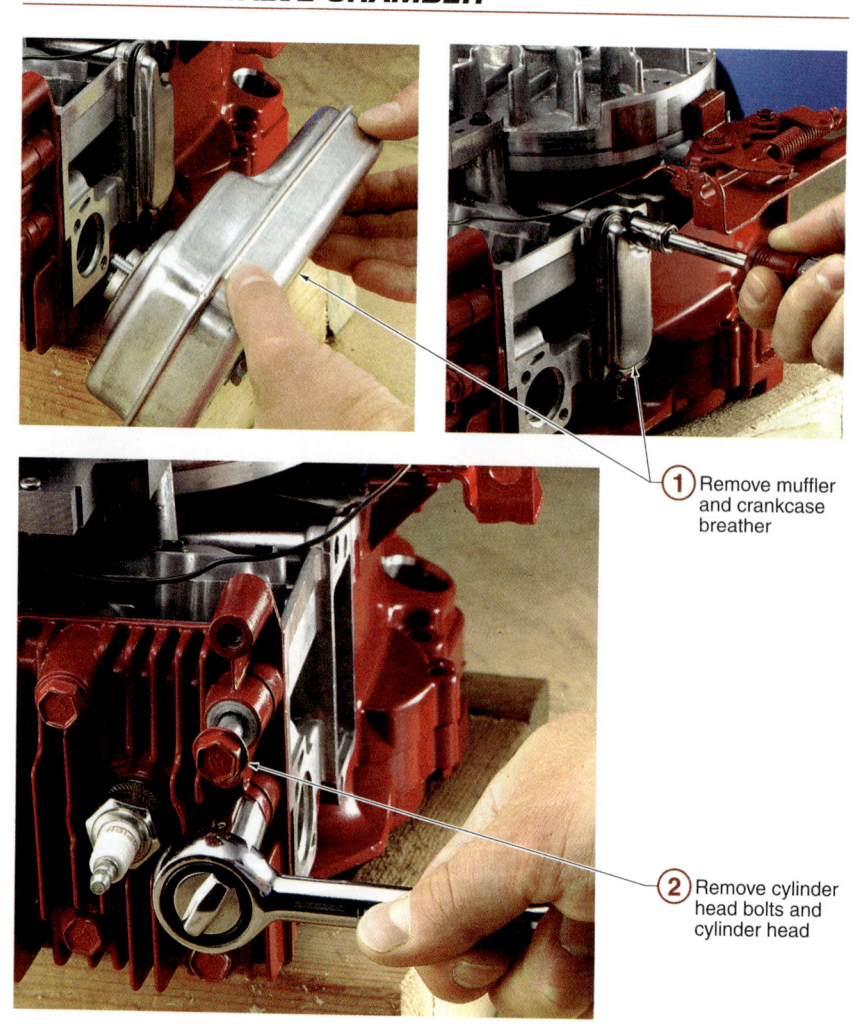

Figure 8-18. The valve chamber on a small engine is accessed by removing the muffler, crankcase breather, and cylinder head.

Chapter 8—Advanced Small Engine Maintenance and Repair Projects 165

To remove the spring (keyhole) retainer of a valve, apply the following procedure:

1. Slip the upper jaw of the valve spring compressor over the top of the valve chamber and the lower jaw between the spring and retainer. If the engine design does not permit the upper jaw to fit over the top of the valve chamber, insert the upper jaw into the chamber on the top of the spring so that the spring is between the jaws of the tool.
2. Rotate the handle on the valve spring compressor clockwise to compress the spring. Slide the retainer off the valve by shifting it with needle-nose pliers so that the large part of the keyhole is directly over the stem. Use needle-nose pliers to remove the retainer from the valve chamber.
3. With the valve spring compressor clamping the spring, remove the tool and spring from the chamber. Slowly crank open the valve spring compressor to release the tension and remove the spring. ***See Figure 8-19.***

Valve Spring Compressors

A valve spring compressor is a required tool for removing valves from an engine. Removing valves without this tool is unsafe and can be difficult. The method for using a valve spring compressor varies depending on the type of valve assembly and the design of the engine block. Some valve assemblies hold the valve springs in place with automotive-type or pin retainers. Others use retainers with keyhole-shaped slots that lock onto the valve stems. Regardless of the type of retainer on the valves, a valve spring compressor allows for proper valve removal and installation.

REMOVING SPRING RETAINERS

Valve spring compressor

① Place spring between jaws of valve spring compressor

② Compress spring and use needle-nose pliers to remove spring retainer

③ Remove spring from valve chamber

Figure 8-19. Spring retainers are removed with a valve spring compressor and needle-nose pliers.

...Project: Servicing Valves...

To remove a valve with an automotive-type or pin retainer, apply the following procedure:

REMOVING VALVES

1. Adjust the jaws of the valve spring compressor until they touch the top and bottom of the valve chamber.
2. Push the tool in until the upper jaw slips over the upper end of the spring. Tighten the jaws to compress the spring.
3. Remove the retainer and lift out the valve, compressor, and spring. *See Figure 8-20.*

① Adjust jaws of valve spring compressor until they touch top and bottom of valve chamber

② Tighten jaws to compress spring

③ Lift out valve, compressor, and spring

Figure 8-20. Valves are removed after compressing the springs and removing the retainers, compressors, and springs.

To inspect a valve, apply the following procedure:

1. Check the valve face for an irregular seating pattern. The pattern around the face should be even with the valve head and of equal thickness around the valve head circumference. Also check for hard deposits. After soaking the parts in solvent for several hours, if necessary, to loosen hardened grit, remove the hard deposits with a wire brush. *Note:* The exhaust valve spring may use a thicker wire than the intake valve spring. Before wiping or cleaning the valves, inspect them carefully.

2. Run a fingernail (or piece of thin, rigid plastic) along the circumference of the valve stem once it has been cleaned. If a ridge exists, the valve stem is worn and should be replaced. The valve guide may also be worn and need to be reworked by a machinist.

3. Measure the thickness of the valve head (valve head margin) with a caliper. Replace the valve if the margin measures less than 1/64″.

4. Examine the surfaces of the valve face and seat. An uneven wear pattern indicates it is time to replace them both or to resurface the seat and replace the valve.

5. Check the valve springs for straightness. Replace any bent springs. *See Figure 8-21.*

Note: Residue on the valves may help identify a specific problem. Gum deposits on the intake valve can cause a reduction in engine performance, often because the engine has been run using old gasoline. Hard deposits on the valves are formed from burning oil.

INSPECTING VALVES

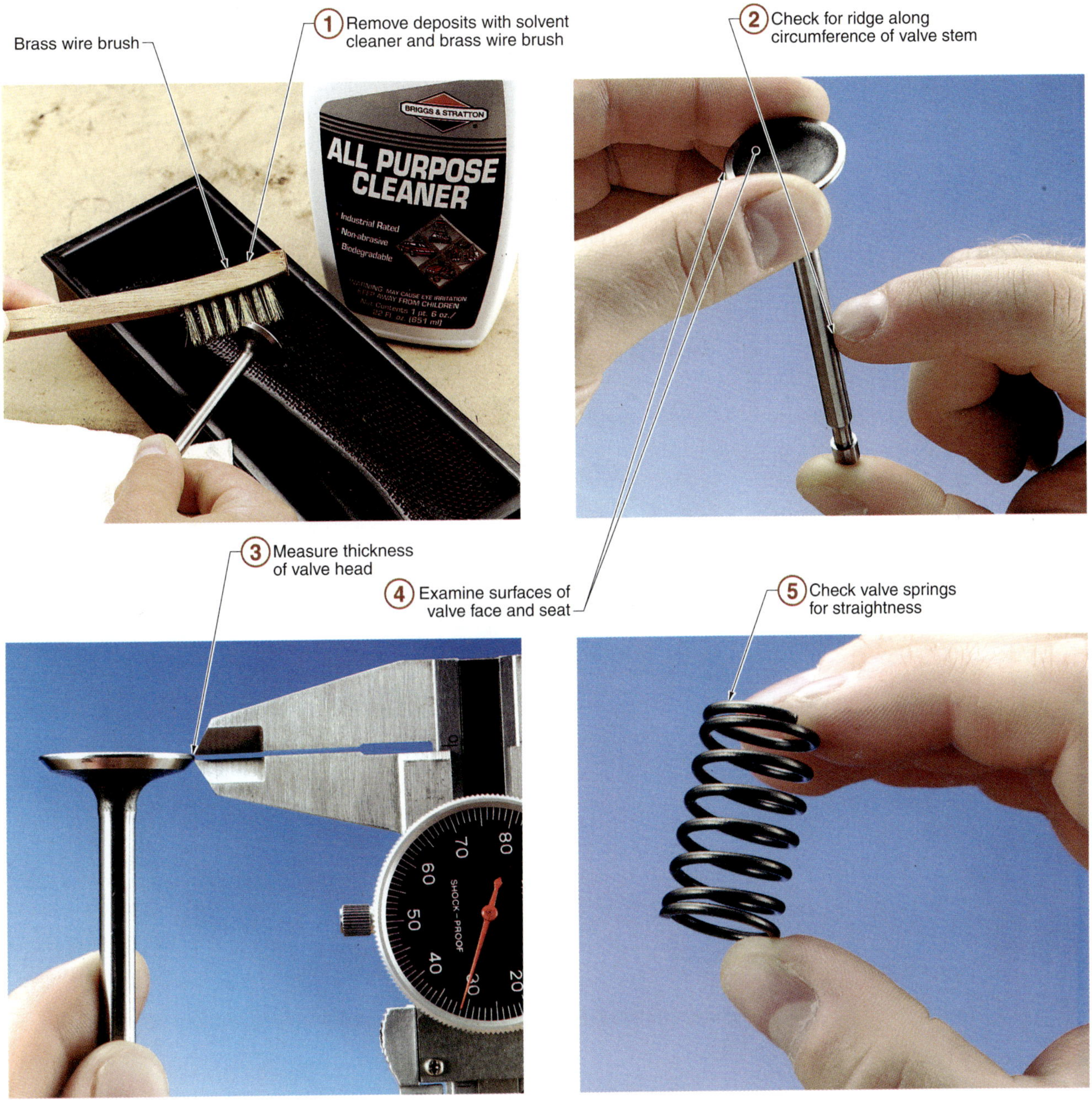

1. Remove deposits with solvent cleaner and brass wire brush
   - Brass wire brush
2. Check for ridge along circumference of valve stem
3. Measure thickness of valve head
4. Examine surfaces of valve face and seat
5. Check valve springs for straightness

Figure 8-21. Valves are inspected to help identify the source of an engine problem.

... Project: Servicing Valves ...

Replacing Valve Seat Inserts

Engines with cast iron cylinders are equipped with exhaust valve seat inserts. When the surfaces of the valve seat inserts are worn and cannot be restored by lapping, an authorized Briggs & Stratton service technician can remove the inserts and install new ones. The intake side must be counterbored to allow the installation of a seat insert.

To lap a valve, apply the following procedure:

1. Apply a small amount of valve lapping compound to the valve face and insert the valve into the valve guide.
2. Wet the end of the lapping tool suction cup, and place it on the valve head. Rotate the valve back and forth between the hands several times. Lift the tool, rotate a quarter turn, and rotate the valve back and forth again. *Note:* Clean the surface frequently and check progress. Lap only enough to create a consistent and even pattern around the valve face.
3. After lapping, clean the valve thoroughly with solvent to ensure all abrasive residue is removed. Any particles that remain can rapidly damage the valve and other engine components. ***See Figure 8-22.***

LAPPING VALVES

① Apply valve lapping compound to valve face, and insert into valve guide
② Use valve lapping tool to rotate valve back and forth several times
③ Clean valves with solvent to remove any abrasives remaining from lapping compound

Figure 8-22. Valve lapping is done to repair minor scoring and pitting of the valve face and seat and to restore the ability of the valve to seal the combustion chamber.

Valve Life

The life of a standard exhaust valve is often shortened due to overheating, which occurs when combustion deposits lodge between the valve seat and the valve face. These deposits prevent the valves from closing and sealing completely. Engines that operate at constant speeds and constant loads for long periods of time, such as generator engines, are more susceptible to overheating and valve problems. Valve life can be extended by using a valve rotator, which turns the exhaust valve slightly on each lift, wiping away the deposits lodged between the valve face and seat. Cobalite® exhaust valves, which have a high resistance to heat and can be used with valve rotators, reduce the occurrence of overheating.

To adjust tappet clearance, apply the following procedure:

1. With each valve installed in its proper guides in the cylinder, rotate the crankshaft (clockwise as viewed from the flywheel end of the crankshaft) to top dead center (TDC). Both valves should be closed. Rotate the crankshaft past TDC until the piston is ¼″ down from the top of the cylinder.
2. Check the clearance between each valve and its tappet, using a feeler gauge.
3. If clearance is insufficient, remove the valve and grind or file the end of the valve stem square to increase the clearance. Check the length frequently, as it is easy to remove too much metal.
4. Once the individual valve parts have been thoroughly cleaned, lubricate the valve stems and guides using valve guide lubricant, ensuring there is no lubricant on the ends of the valve stems or tappets.

To reinstall a valve with a spring (keyhole) retainer, apply the following procedure:

1. Compress the keyhole retainer and spring with a valve spring compressor (the large hole should face the opening in the tool) until the spring is completely compressed.
2. Brush the valve stem with valve stem lubricant.
3. Insert the compressed spring and retainer into the valve chamber.
4. Insert the valve stem through the large slot in the retainer. Push down and in on the valve compressor until the retainer bottoms out on the valve stem shoulder. Release spring tension.
5. Reinstall the crankcase breather and other components. *See Figure 8-23.*

REINSTALLING VALVES WITH SPRING RETAINERS

Figure 8-23. Valves with spring retainers are reinstalled using valve spring compressors.

...Project: Servicing Valves...

WARNING: Always wear eye protection, such as safety glasses or goggles, when working with valves and valve springs for protection from flying springs or other parts.

To reinstall a valve with an automotive-type or pin retainer, apply the following procedure:

1. Place the valve spring into the valve spring compressor. Rotate the handle of the tool until the spring is completely compressed.
2. Insert the compressed spring into the valve chamber.
3. Brush the valve stem with valve stem lubricant, and lower the valve stem through the spring.
4. Hold the spring toward the top of the chamber and the valve in the closed position.
5. If pins are used, insert each pin with needle-nose pliers. If automotive-type retainers are used, place the retainers in the valve stem groove.
6. Lower the spring until the retainer fits around the automotive-type or pin retainer, and then pull out the valve spring compressor.
7. Reinstall the crankcase breather and other components.

To remove Briggs & Stratton Intek® 6 HP single-cylinder overhead valves (OHVs), which do not require the use of a valve spring compressor, apply the following procedure:

1. Remove the air cleaner assembly, fuel tank, oil fill tube, blower housing, rewind starter, muffler guard, muffler, carburetor, and any other parts that block access to the cylinder head.
2. Remove the screws from the valve cover using a socket wrench or nut driver. Remove the valve cover, breather valve assembly (if equipped), and any gaskets.
3. Remove the rocker arm bolts with a socket wrench or nut driver, and then remove the rocker arms and push rods.
4. Remove the valve caps (if equipped).
5. Use both thumbs to press the spring retainer and valve spring over one of the valves. With the valve spring compressed, remove the valve spring retainer. *Note:* If the engine has a keyhole retainer, line up the large slot in the retainer with the valve stem and release the spring slowly so that the stem slips through the large slot. Repeat this step for the other valve.
6. Remove the push rod guide bolts and push rod guide.
7. Remove the cylinder head bolts and cylinder head (by hand-rocking it). If necessary, loosen the cylinder head by striking it with a nylon-face hammer. Never pry it loose with a screwdriver, as this may damage the head.
8. Remove and inspect the valves, guides, and seats. *Note:* The intake and exhaust valves often are composed of different steel alloys and may be different colors. ***See Figure 8-24.***

REMOVING OVERHEAD VALVES

1. Remove fuel tank and other parts that block access to valves
2. Remove valve cover
3. Remove rocker arms and push rods
4. Remove valve caps
5. Remove valve spring retainer
6. Remove push rod guide bolts and push rod guide
7. Remove cylinder head bolts and cylinder head
8. Remove and inspect valves, guides, and seat

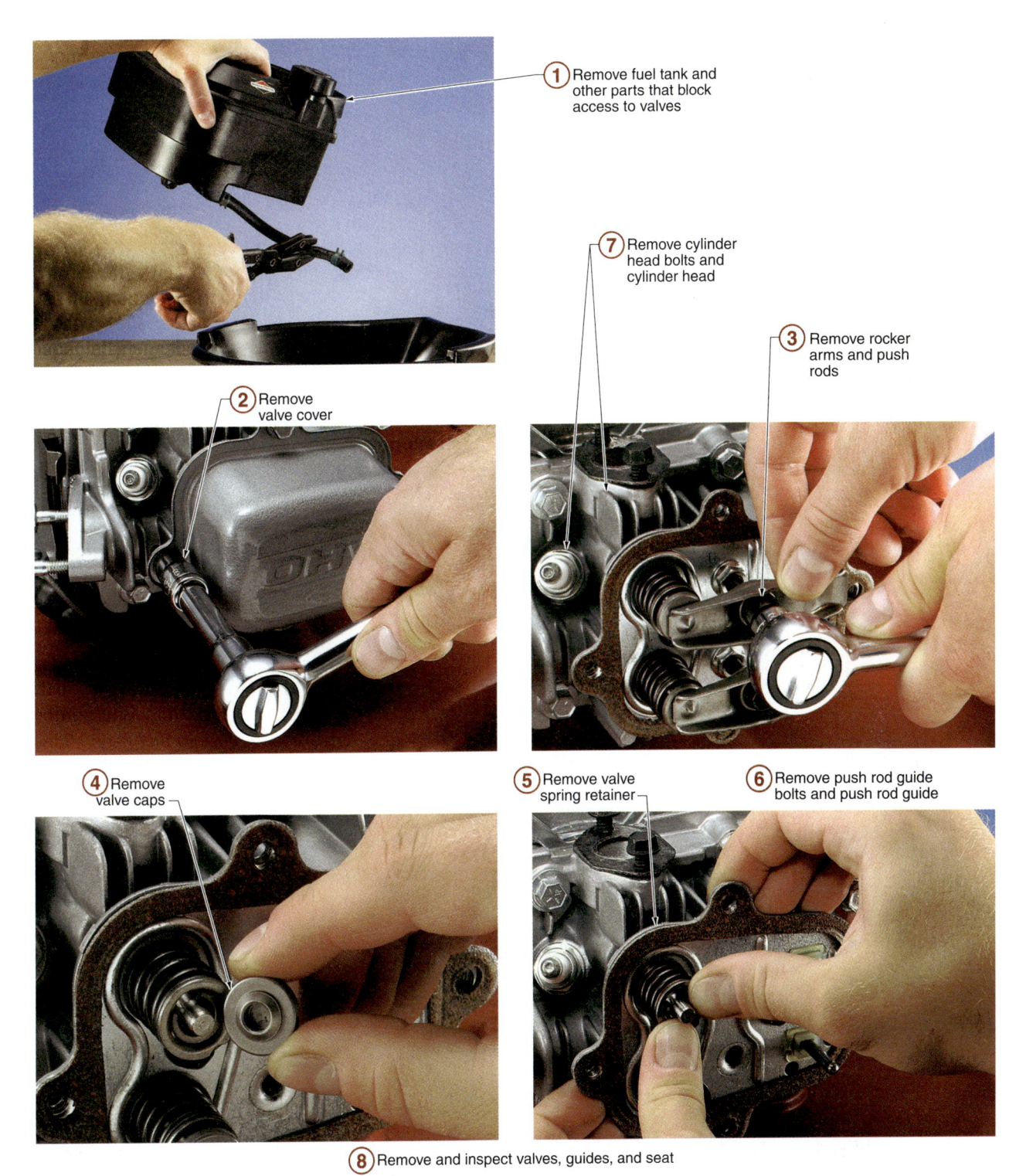

Figure 8-24. The parts of an overhead valve cylinder head may differ, depending on the engine model and manufacturer.

...Project: Servicing Valves...

To reinstall overhead valves, apply the following procedure:

1. Ensure the valve stems and guides are free of debris and burrs. Lightly coat the valve stems with valve guide lubricant or engine oil, and insert them in the cylinder head, taking care to place the correct valve in each valve guide.
2. Place the push rod guide on the cylinder head and attach with mounting bolts using a torque wrench.
3. Coat the threads of the rocker arm stud with a hardening sealant. Install the rocker arm studs using a socket wrench. Consult an authorized Briggs & Stratton dealer for the proper torque settings for the mounting bolts and studs.
4. Lubricate the inside diameter of each valve stem seal (if equipped) with engine oil. Install the seals on the valve stems by pressing them into place.
5. Install a valve spring and retainer over each stem. Use both thumbs to compress the spring until the valve stem extends through the large end of the keyhole slot. Check that the retainer is fully engaged in the valve stem groove. Repeat this step for the other valve.
6. Coat the threads of the cylinder head bolts with valve guide lubricant. Install a new cylinder head gasket on the cylinder. Insert the bolts in the head, and position the cylinder head on the cylinder.
7. Tighten the cylinder head bolts in increments using a torque wrench. Turn each bolt a few turns, and proceed to the next bolt until each bolt is snug. For final tightening, proceed in increments of about one-third the final torque. Consult the owner's/operator's manual for final torque specifications. Uneven tightening can warp the cylinder head.
8. Install push rods through the push rod guides and into the tappets.
9. Install the caps on the ends of the valves, and wipe away any lubricant. Install the rocker arm assemblies while holding the rocker arms against the valve cap and push rod.
10. Rotate the flywheel at least two revolutions to verify that the push rods operate the rocker arms. *See Figure 8-25.*

REINSTALLING OVERHEAD VALVES

① Coat valve stems with lubricant and insert in cylinder head

② Place push rod guide on cylinder head, and attach mounting bolts with torque wrench

③ Install rocker arm studs

④ Install system valve stem seals

⑤ Install spring and retainer over each valve stem

⑥ Install head gasket and cylinder head

⑦ Tighten cylinder head bolts with torque wrench

⑧ Install push rods through push rod guides and into tappets

⑨ Install rocker arm assemblies

⑩ Rotate flywheel to check push rod operation

Figure 8-25. Overhead valves are reinstalled using a torque wrench and socket wrench or nut driver set.

...Project: Servicing Valves...

To adjust overhead valves, apply the following procedure:

1. Release the brake spring, and rotate the flywheel to close both valves.
2. Insert a narrow screwdriver into the spark plug hole and touch the piston. Rotate the flywheel clockwise past TDC until the piston has moved down ¼″. Use the screwdriver to gauge the range of motion of the piston.
3. Check the valve clearance by placing a feeler gauge between the valve head and the rocker arm. *Note:* Clearances differ for the two valves and typically range from 0.002″ to 0.004″ and from 0.005″ to 0.007″. Consult an authorized Briggs & Stratton dealer for the proper valve clearances for a specific engine model.
4. Adjust the valve clearances as required by rotating the rocker screw. Once adjustments are completed, tighten the rocker nut.
5. Install the valve cover using new gaskets as required. Verify that the cover is secure. *See Figure 8-26.*

Chapter 8 Quick Quiz®

ADJUSTING OVERHEAD VALVES

① Release brake spring and rotate flywheel to close valves

② Use screwdriver in spark plug hole to gauge range of motion of piston

③ Check valve clearance with feeler gauge

④ Adjust valve clearances by rotating rocker screw

⑤ Install valve cover and new gaskets

Figure 8-26. Overhead valves must be adjusted for proper operation once they are reinstalled.

Appendix

OPE Maintenance Log — 176

Additional Tools Parts and Supplies — 177

Industry and Standards Organizations — 178

Engine Classification — 179

Metric Prefixes — 180

Metric Conversions — 180

Measures — 181

Fuel/Oil Mix — 181

Air Density Effects — 182

Failure Analysis — 183

Governor System Troubleshooting — 186

English System — 188

Metric System — 189

Warranty Claim — 190

Safety Color Coding — 191

Exhaust Evacuation System — 191

Hazardous Material Container Labeling — 192

Hazardous Materials — 193

Horsepower to Torque Conversion — 194

OPE MAINTENANCE LOG

Date Performed	Maintenance Task	Every 5 hours	Every 25 Hours or Every Season	Every 50 Hours or Every Season	Every 100 Hours or Every Season	Every 100-300 Hours
	Check oil level	X				
	Change oil*			X		
	Replace oil-foam® element or optional pre-cleaner†		X			
	Replace air-cleaner cartridge if no pre-cleaner†		X			
	Replace air-cleaner cartridge if equipped with pre-cleaner†				X	
	Clean cooling system†				X	
	Inspect spark-arrestor (if so equipped)			X		
	Replace in-line fuel filter (optional accessory)			X		
	Replace spark plug				X	
	Clean combustion chamber deposits					X

* Change oil after first 5 hours of use, then every 50 hours or every season
† Clean more often under dusty conditions or when airborne debris is present; replace air-cleaner parts if very dirty

ADDITIONAL TOOLS, PARTS, AND SUPPLIES

Common Replacement Parts

- Air cleaner
- Carburetor
- Connecting rod
- Fuel filter
- Muffler
- Oil filter
- Piston
- Piston rings
- Spark plug
- Starter rope and grip
- Valve
- Valve retainer

Lubricants and Cleaners

- 2-cycle engine oil
- 4-use lubricant
- Battery cleaner
- Battery terminal protector
- Carburetor/choke cleaner
- Fogging oil
- Gasoilne additive
- Grease gun
- Heavy-duty degreaser
- Heavy-duty silicone
- Lawn mower oil
- Moly/graphite grease
- Penetrating oil
- Valve guide lubricant
- Valve lapping compound
- White lithium grease

Specialty Tools

- Brake adjustment gauge: For setting band brakes
- Carburetor jet screwdrivers: For removing and installing carburetor jets
- C-ring installation tool: For installing starter motor C-ring
- C-ring removal tool: For removing C-ring on starter motor
- Cylinder leakdown tester: For testing sealing capabilities of compression components
- DC shunt: For measuring current draw of DC motors and output of regulated alternators
- Flywheel holder: For removing and installing flywheel
- Leakdown tester clamp: For holding leakdown tester in place
- Piston ring compressor: For compressing piston rings during assembly
- Piston ring expander: For removing and installing piston rings
- Plug gauge: For checking wear on valve guides
- Starter clutch wrench: For removing and torquing rewind starter clutch
- Telescoping gauge: For measuring inside diameters of cylinders
- Torque wrench: For tightening bolts to specific inch pounds of torque
- Valve lapping tool: For resurfacing valve faces and seats

INDUSTRY AND STANDARDS ORGANIZATIONS

CPSC
Consumer Products Safety Commission
4330 East West Highway
Bethesda, MD 20814
www.cpsc.gov

DOT
Department of Transportation
1200 New Jersey Avenue SE
Washington, DC 20590
www.dot.gov

U.S. EPA
U.S. Environmental Protection Agency
Ariel Rios Building
1200 Pennsylvania Avenue NW
Washington, DC 20460
www.epa.gov

NIOSH
National Institute for Occupational
Safety and Health
1600 Clifton Road
Atlanta, GA 30333
www.cdc.gov/niosh

OSHA
Occupational Safety and
Health Administration
200 Constitution Avenue NW
Washington, DC 20210
www.osha.gov

Government Agencies

ANSI
American National Standards Institute
1819 L Street NW, 11th Floor
Washington, DC 20036
www.ansi.org

CSA
Canadian Standards Association
5060 Spectrum Way, Suite 100
Mississauga, ON L4W 5N6
www.csa.ca

ISO
International Organization for Standardization
Case Postale 56 CH - 1211
Geneve 20 Switzerland
www.iso.org

Standards Organizations

SAE
Society of Automotive Engineers
400 Commonwealth Drive
Warrendale, PA 15096
www.sae.org

ASABE
American Society of Agricultural
and Biological Engineers
2950 Niles Road
St. Joseph, MI 49085
www.asbe.org

ASTM International
American Society for Testing and Materials
100 Barr Harbor Drive
West Conshohocken, PA 19428
www.astm.org

Technical Societies

NFPA
National Fire Protection Association
1 Batterymarch Park
Quincy, MA 02269
www.nfpa.org

UL
Underwriters Laboratories, Inc.
333 Pfingston Road
Northbrook, IL 60062
www.ul.com

Trade Organizations

API
American Petroleum Institute
1220 L Street NW
Washington, DC 20005
www.api.org

OPEI
Outdoor Power Equipment Institute
341 South Patrick Street
Old Town Alexandria, VA 22314
www.opei.org

Private Organizations

EETC
Equipment & Engine Training Council
PO Box 1078
N59 W39556 Laketon Avenue
Oconomowoc, WI 53066
www.eetc.org

OPEESA
Outdoor Power Equipment and
Engine Service Association
37 Pratt Street
Essex, CT 06426
www.opeesa.com

AED
AED Foundation
Associated Equipment Distributors, Inc.
600 Hunter Drive, Suite 200
Oak Brook, IL 60523
www.aednet.org

Training-Related Associations

SkillsUSA
SkillsUSA
14001 SkillsUSA Way
Leesburg, VA 20176
www.skillsusa.org

FFA
National FFA Organization
PO Box 68960, 6060 FFA Drive
Indianapolis, IN 46268
www.ffa.org

Student-Related Organizations

Appendix 179

ENGINE CLASSIFICATION

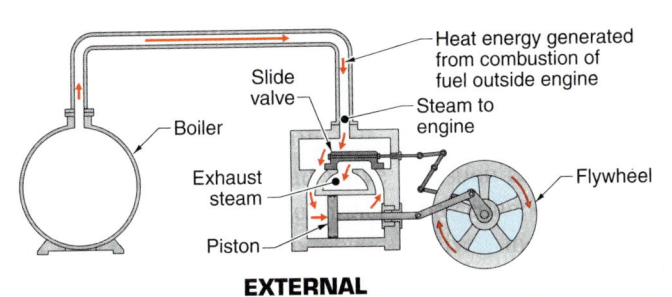

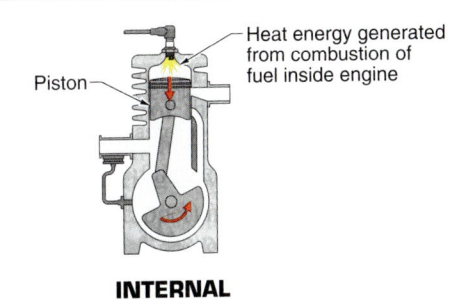

EXTERNAL — Boiler, Slide valve, Steam to engine, Heat energy generated from combustion of fuel outside engine, Exhaust steam, Piston, Flywheel

INTERNAL — Piston, Heat energy generated from combustion of fuel inside engine

Combustion

SPARK (GASOLINE ENGINE) — Spark plug, Air-fuel mixture in combustion chamber, Power stroke, Piston

COMPRESSION (DIESEL ENGINE) — Injector, Compressed air, Air, Power stroke

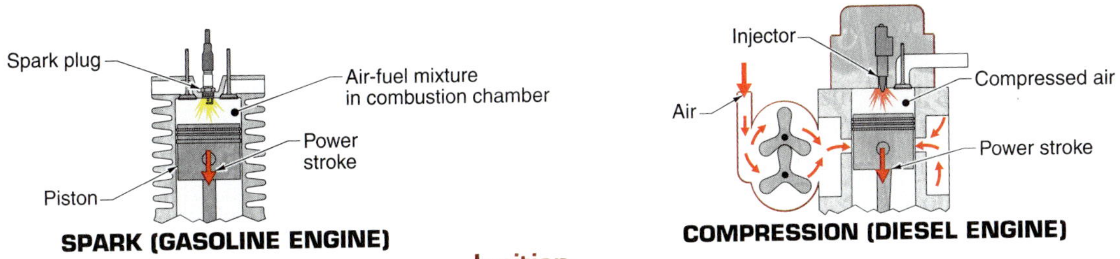

Ignition

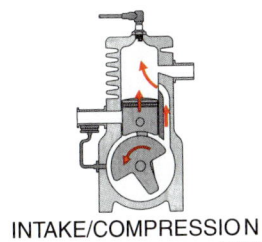

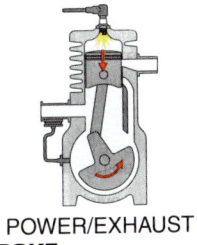

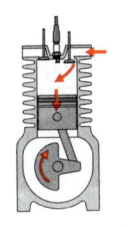

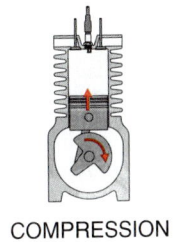

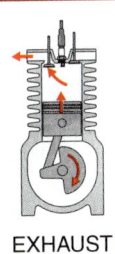

TWO-STROKE — INTAKE/COMPRESSION, POWER/EXHAUST

FOUR-STROKE — INTAKE, COMPRESSION, POWER, EXHAUST

Number of strokes

ORIENTATION — VERTICAL, HORIZONTAL, SLANTED

CONFIGURATION — V, HORIZONTALLY-OPPOSED, IN-LINE

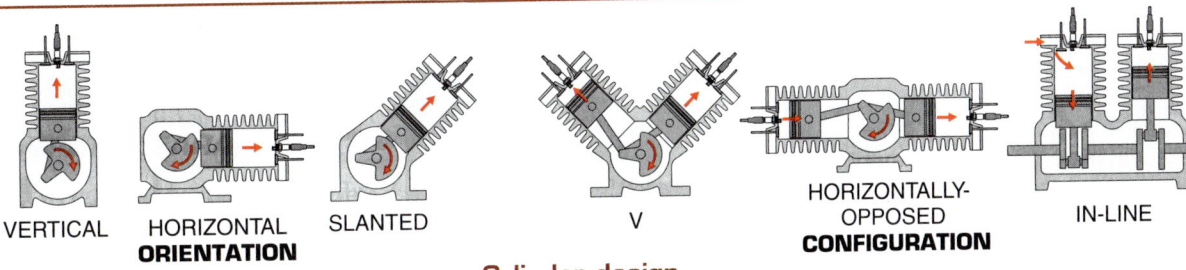

Cylinder design

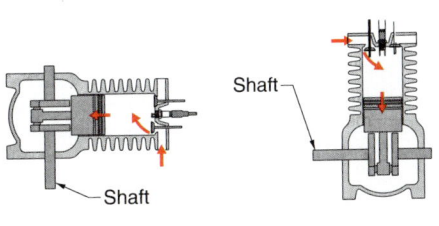

Shaft orientation — VERTICAL (Shaft), HORIZONTAL (Shaft)

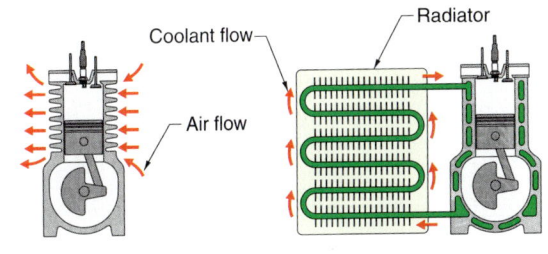

Cooling system — AIR-COOLED, LIQUID-COOLED (Coolant flow, Radiator, Air flow)

METRIC PREFIXES

Multiples and Submultiples	Prefixes	Symbols	Meaning
$1,000,000,000,000 = 10^{12}$	tera	T	trillion
$1,000,000,000 = 10^{9}$	giga	G	billion
$1,000,000 = 10^{6}$	mega	M	million
$1000 = 10^{3}$	kilo	k	thousand
$100 = 10^{2}$	hecto	h	hundred
$10 = 10^{1}$	deka	d	ten
Unit $1 = 10^{0}$			
$.1 = 10^{-1}$	deci	d	tenth
$.01 = 10^{-2}$	centi	c	hundredth
$.001 = 10^{-3}$	milli	m	thousandth
$.000001 = 10^{-6}$	micro	µ	millionth
$.000000001 = 10^{-9}$	nano	n	billionth
$.000000000001 = 10^{-12}$	pico	p	trillionth

METRIC CONVERSIONS

Initial Units	Final Units										
	kilo	hecto	kilo	hecto	deka	base	deci	centi	milli	micro	nano	pico
giga		3R	6R	7R	8R	9R	10R	11R	12R	15R	18R	21R
mega	3L		3R	4R	5R	6R	7R	8R	9R	12R	15R	18R
kilo	6L	3L		1R	2R	3R	4R	5R	6R	9R	12R	15R
hecto	7L	4L	1L		1R	2R	3R	4R	5R	8R	11R	14R
deka	8L	5L	2L	1L		1R	2R	3R	4R	7R	10R	13R
base unit	9L	6L	3L	2L	1L		1R	2R	3R	6R	9R	12R
deci	10L	7L	4L	3L	2L	1L		1R	2R	5R	8R	11R
centi	11L	8L	5L	4L	3L	2L	1L		1R	4R	7R	10R
milli	12L	9L	6L	5L	4L	3L	2L	1L		3R	6R	9R
micro	15L	12L	9L	8L	7L	6L	5L	4L	3L		3R	6R
nano	18L	15L	12L	11L	10L	9L	8L	7L	6L	3L		3R
pico	21L	18L	15L	14L	13L	12L	11L	10L	9L	6L	3L	

MEASURES

Linear
12 inches = 1 foot
3 feet = 1 yard
36 inches = 1 yard
5.5 yards = 1 rod
16.5 feet = 1 rod
40 rods = 1 furlong
660 feet = 1 furlong
8 furlongs = 1 mile
320 rods = 1 mile
1760 yards = 1 mile
5280 feet = 1 mile

Surveyor's Linear
7.92 inches = 1 link
16.5 feet = 1 rod
25 links = 1 rod
4 rods = 1 chain
66 feet = 1 chain
100 links = 1 chain
80 chains = 1 mile

Dry
2 pints = 1 quart
4 quarts = 1 gallon
2 gallons = 1 peck
8 quarts = 1 peck
4 pecks = 1 bushel

Time
60 seconds = 1 minute
60 minutes = 1 hour
24 hours = 1 day
7 days = 1 week
52 weeks = 1 year
365.26 days = 1 year

Avoirdupois
437.5 grains = 1 ounce
16 ounces = 1 pound
100 pounds = 1 hundredweight
1000 pounds = 1 kip
2 kips = 1 ton
2000 pounds = 1 ton
2240 pounds = 1 long ton

Surveyor's Square
625 square links = 1 square rod
16 square rods = 1 square chain
10 square chains = 1 acre
640 acres = 1 square mile
1 square mile = 1 section
36 square miles = 1 township
36 sections = 1 township

Liquid
4 gills = 1 pint
2 pints = 1 quart
57.75 cubic inches = 1 quart
4 quarts = 1 gallon
231 cubic inches = 1 gallon
31.5 gallons = 1 barrel

Square
144 square inches = 1 square foot
9 square feet = 1 square yard
1296 square inches = 1 square yard
30.25 square yards = 1 square rod
160 square rods = 1 acre
4840 square yards = 1 acre
43,560 square feet = 1 acre
640 acres = 1 square mile

Cubic
7.48 gallons = 1 cubic foot
1728 cubic inches = 1 cubic foot
27 cubic feet = 1 cubic yard
202 gallons = 1 cubic yard
128 cubic feet = 1 cord

Angular
60 seconds = 1 minute
60 minutes = 1 degree
57.3 degrees = 1 radian
180 degrees = π radians
360 degrees = 2π radians

FUEL/OIL MIX

	US Gallons		Imperial Gallons		Metric			US Gallons		Imperial Gallons		Metric	
	Fuel*	Oil†	Fuel*	Oil†	Petrol‡	Oil‡		Fuel*	Oil†	Fuel*	Oil†	Petrol‡	Oil‡
16 : 1	1	8	1	10	4	0.250	32 : 1	1	4	1	5	4	0.125
	3	24	3	30	12	0.750		3	12	3	15	12	0.375
	5	40	5	50	20	1.250		5	20	5	25	20	0.625
	6	48	6	60	24	1.500		6	24	6	30	24	0.750
24 : 1	1	5.33	1	6.4	4	0.160	50 : 1	1	2.5	1	3	4	0.080
	3	16	3	19.2	12	0.470		3	8.0	3	9	12	0.240
	5	26.66	5	32.0	20	0.790		5	13.0	5	15	20	0.400
	6	32	6	38.4	24	0.940		6	15.5	6	18.5	24	0.480

* in gal.
† in oz
‡ in l

182 SMALL ENGINE AND EQUIPMENT MAINTENANCE

AIR DENSITY EFFECTS

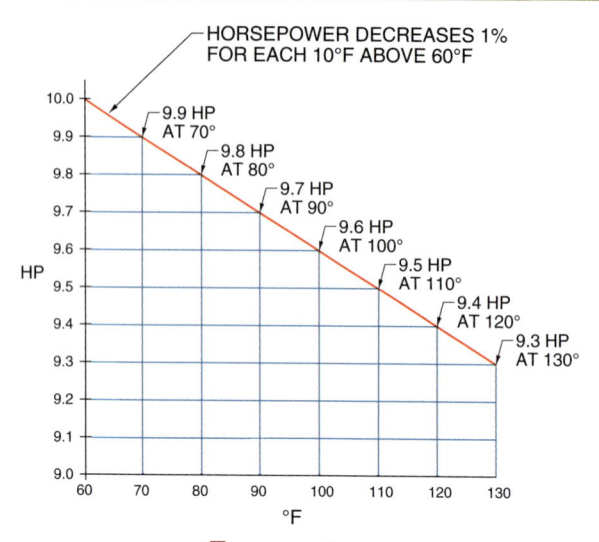

Temperature

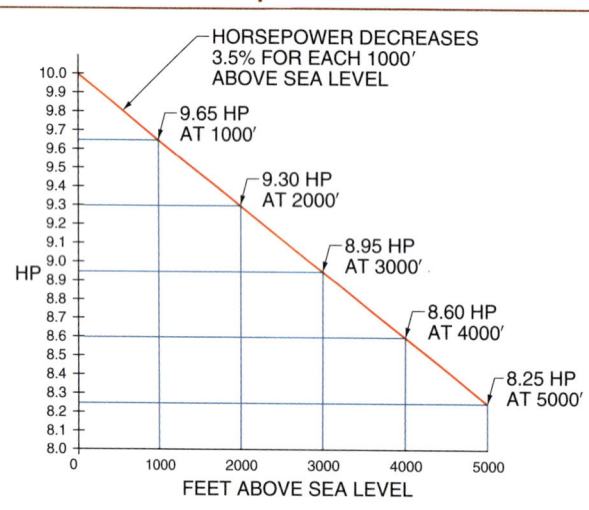

HORSEPOWER DECREASES 1% FOR EACH 10°F ABOVE 60°F

- 9.9 HP AT 70°
- 9.8 HP AT 80°
- 9.7 HP AT 90°
- 9.6 HP AT 100°
- 9.5 HP AT 110°
- 9.4 HP AT 120°
- 9.3 HP AT 130°

Altitude

HORSEPOWER DECREASES 3.5% FOR EACH 1000' ABOVE SEA LEVEL

- 9.65 HP AT 1000'
- 9.30 HP AT 2000'
- 8.95 HP AT 3000'
- 8.60 HP AT 4000'
- 8.25 HP AT 5000'

Appendix 183

FAILURE ANALYSIS...

Abrasive ingestion, upper end

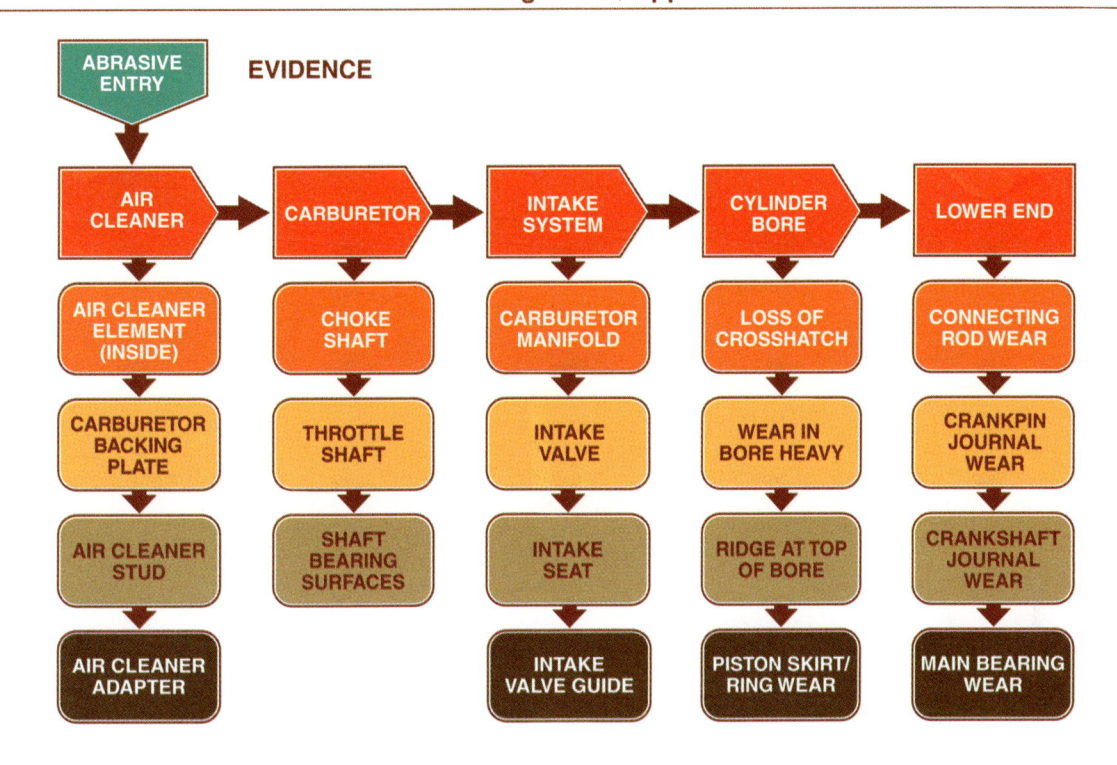

ABRASIVE ENTRY → **EVIDENCE**

- **AIR CLEANER** → AIR CLEANER ELEMENT (INSIDE) → CARBURETOR BACKING PLATE → AIR CLEANER STUD → AIR CLEANER ADAPTER
- **CARBURETOR** → CHOKE SHAFT → THROTTLE SHAFT → SHAFT BEARING SURFACES
- **INTAKE SYSTEM** → CARBURETOR MANIFOLD → INTAKE VALVE → INTAKE SEAT → INTAKE VALVE GUIDE
- **CYLINDER BORE** → LOSS OF CROSSHATCH → WEAR IN BORE HEAVY → RIDGE AT TOP OF BORE → PISTON SKIRT/RING WEAR
- **LOWER END** → CONNECTING ROD WEAR → CRANKPIN JOURNAL WEAR → CRANKSHAFT JOURNAL WEAR → MAIN BEARING WEAR

Abrasive ingestion, lower end

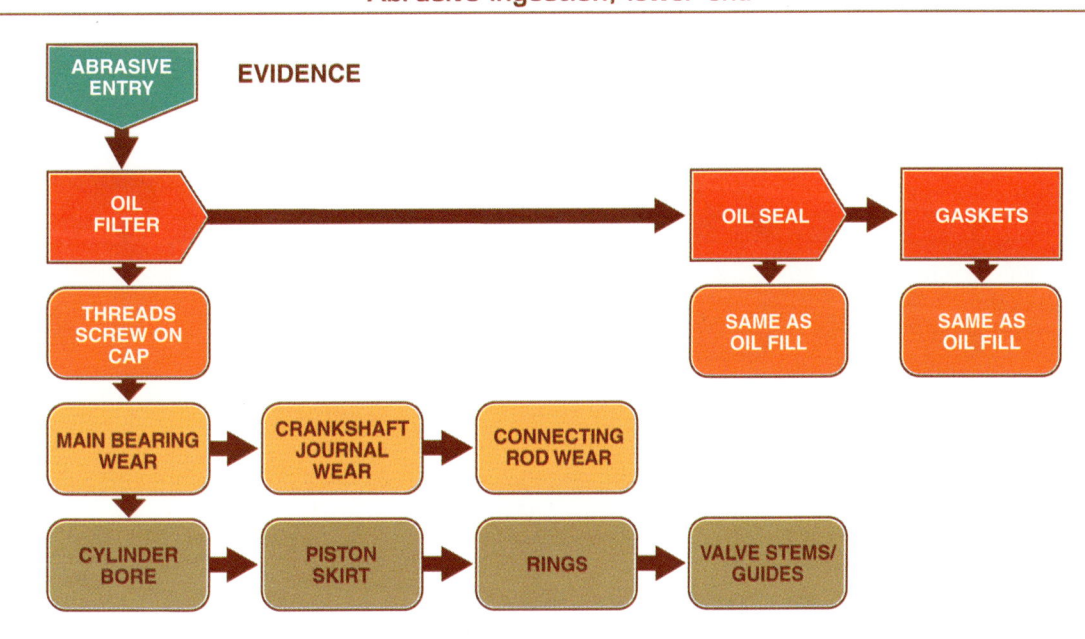

ABRASIVE ENTRY → **EVIDENCE**

- **OIL FILTER** → THREADS SCREW ON CAP → MAIN BEARING WEAR → CYLINDER BORE
  - MAIN BEARING WEAR → CRANKSHAFT JOURNAL WEAR → CONNECTING ROD WEAR
  - CYLINDER BORE → PISTON SKIRT → RINGS → VALVE STEMS/GUIDES
- **OIL SEAL** → SAME AS OIL FILL
- **GASKETS** → SAME AS OIL FILL

184 SMALL ENGINE AND EQUIPMENT MAINTENANCE

...FAILURE ANALYSIS...

Insufficient lubrication

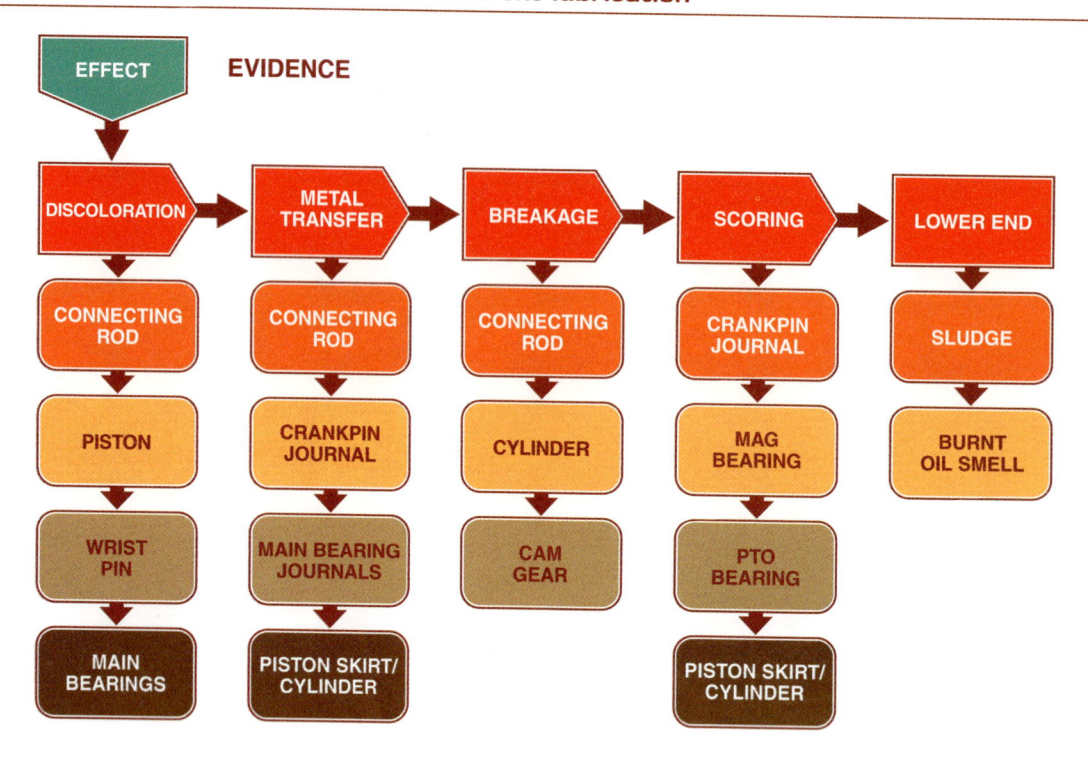

EFFECT → **EVIDENCE**

- DISCOLORATION → CONNECTING ROD → PISTON → WRIST PIN → MAIN BEARINGS
- METAL TRANSFER → CONNECTING ROD → CRANKPIN JOURNAL → MAIN BEARING JOURNALS → PISTON SKIRT/CYLINDER
- BREAKAGE → CONNECTING ROD → CYLINDER → CAM GEAR
- SCORING → CRANKPIN JOURNAL → MAG BEARING → PTO BEARING → PISTON SKIRT/CYLINDER
- LOWER END → SLUDGE → BURNT OIL SMELL

Overheating

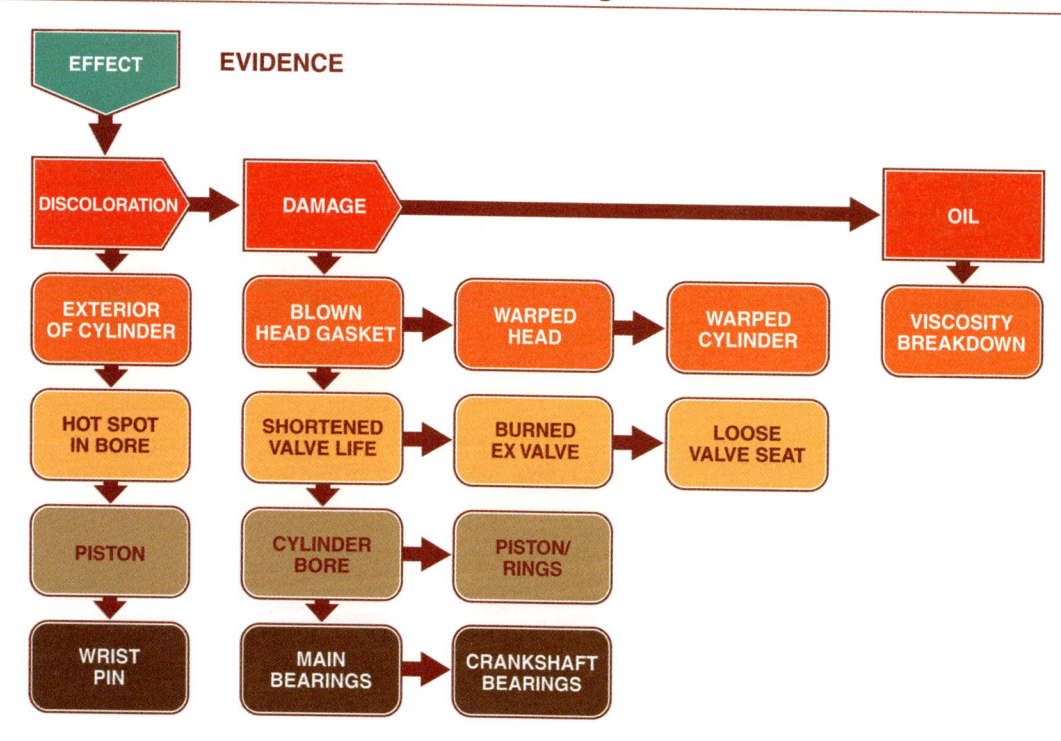

EFFECT → **EVIDENCE**

- DISCOLORATION → EXTERIOR OF CYLINDER → HOT SPOT IN BORE → PISTON → WRIST PIN
- DAMAGE → BLOWN HEAD GASKET → WARPED HEAD → WARPED CYLINDER
- SHORTENED VALVE LIFE → BURNED EX VALVE → LOOSE VALVE SEAT
- CYLINDER BORE → PISTON/RINGS
- MAIN BEARINGS → CRANKSHAFT BEARINGS
- OIL → VISCOSITY BREAKDOWN

...FAILURE ANALYSIS

Overspeeding

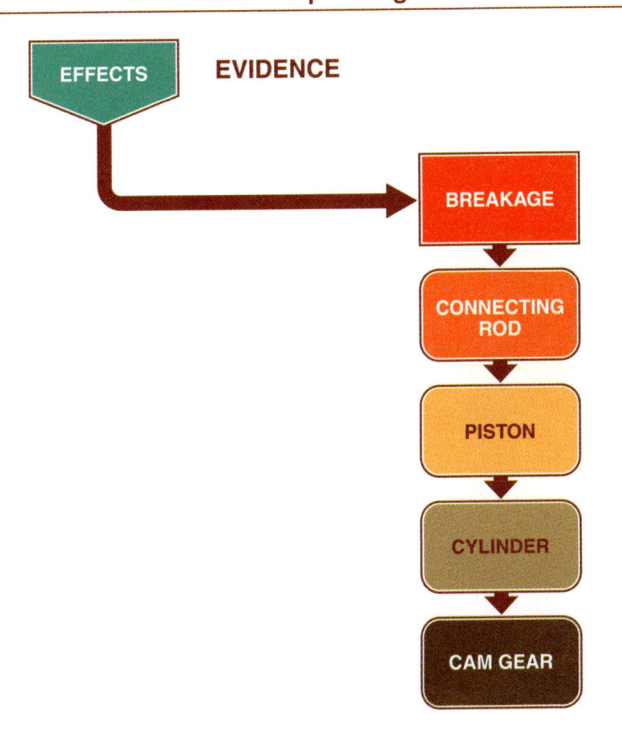

EFFECTS → **EVIDENCE**

- BREAKAGE
- CONNECTING ROD
- PISTON
- CYLINDER
- CAM GEAR

Breakage, vibration

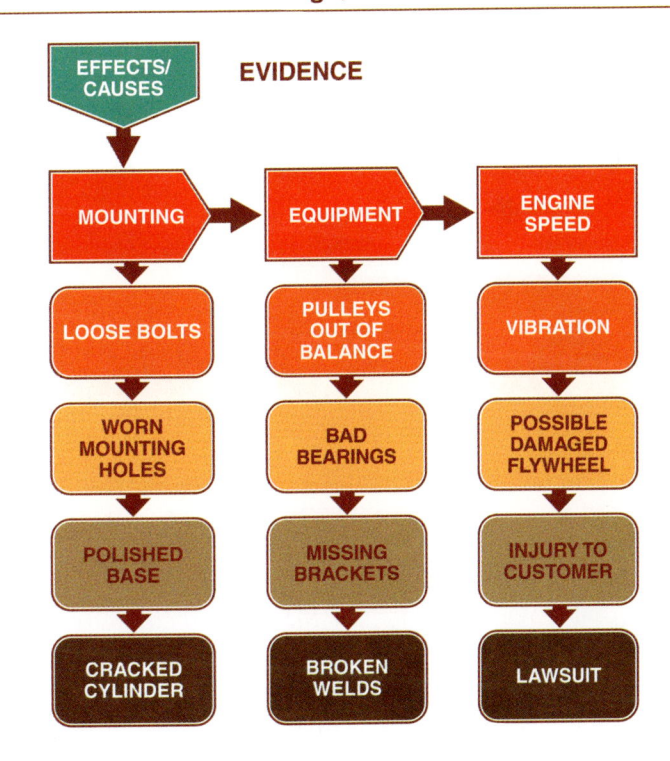

EFFECTS/CAUSES → **EVIDENCE**

MOUNTING	EQUIPMENT	ENGINE SPEED
LOOSE BOLTS	PULLEYS OUT OF BALANCE	VIBRATION
WORN MOUNTING HOLES	BAD BEARINGS	POSSIBLE DAMAGED FLYWHEEL
POLISHED BASE	MISSING BRACKETS	INJURY TO CUSTOMER
CRACKED CYLINDER	BROKEN WELDS	LAWSUIT

186 SMALL ENGINE AND EQUIPMENT MAINTENANCE

GOVERNOR SYSTEM TROUBLESHOOTING...

Pneumatic governor system - overspeeding

- **CHECK COOLING SYSTEM** — IF OK → **REMOVE DIRT AND DEBRIS FROM OUTSIDE AND INSIDE OF BLOWER HOUSING** → **TEST RUN ENGINE**
- **CHECK FLYWHEEL FINS** — IF OK → **REPLACE FLYWHEEL IF FINS ARE MISSING OR BROKEN** → **TEST RUN ENGINE**
- **CHECK FOR UNEVEN OR BINDING RESISTANCE IN GOVERNOR BLADE TRAVEL FROM STOP TO STOP** — IF OK → **IDENTIFY RESISTANCE AND REPAIR** → **TEST RUN ENGINE**
- **INSPECT GOVERNOR SPRING AND CONTROLS FOR DAMAGE OR IMPROPER INSTALLATION** → **REPLACE GOVERNOR SPRING OR CONTROLS** → **TEST RUN ENGINE**

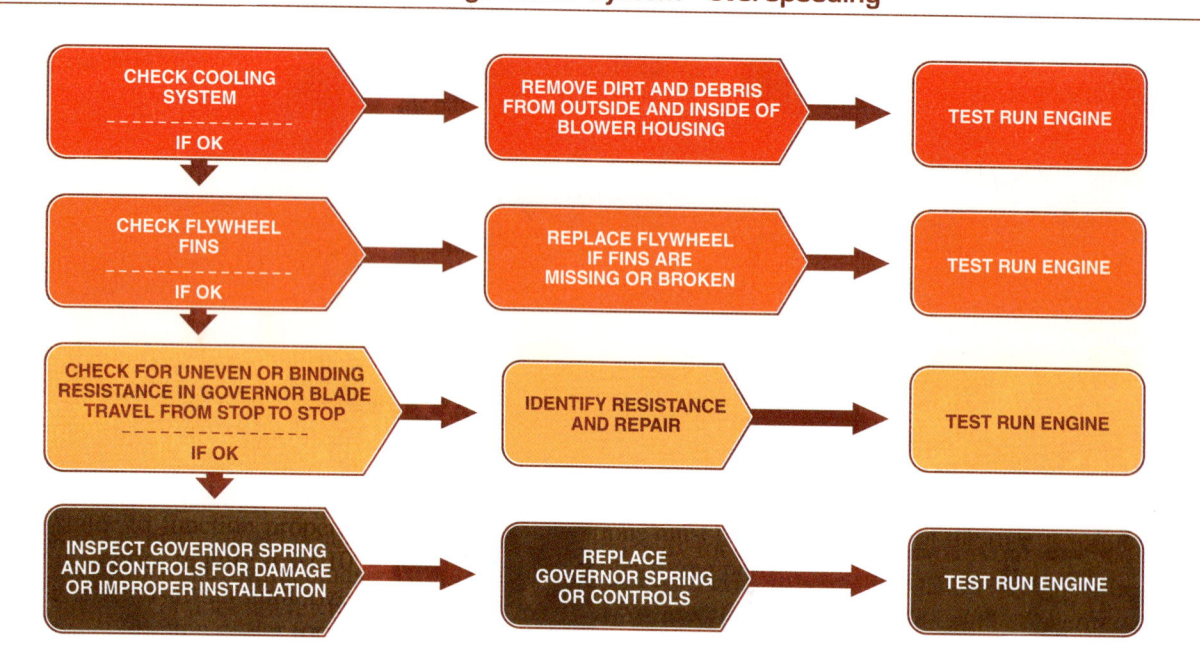

Mechanical governor system - overspeeding

- **PERFORM THE STATIC GOVERNOR ADJUSTMENT PROCEDURE** → **TEST RUN ENGINE** — STILL OVERSPEEDING
- **ADJUST GOVERNOR SPRING TENSION** → **TEST RUN ENGINE** — STILL OVERSPEEDING
- **DISCONNECT GOVERNOR SPRING. CHECK FOR BINDING OR UNEVEN RESISTANCE IN GOVERNOR ARM TRAVEL FROM STOP TO STOP** — NO RESISTANCE → **IDENTIFY RESISTANCE AND REPAIR** → **TEST RUN ENGINE**
- **MOVE THROTTLE PLATE TO WOT POSITION AND START ENGINE IF ENGINE CONTINUES TO OVERSPEED FROM STOP TO STOP** — IF ENGINE RETURNS TO IDLE → **REMOVE CRANKCASE COVER OR SUMP AND INSPECT OR REPLACE INTERNAL GOVERNOR MECHANISM** → **TEST RUN ENGINE**
- **INSPECT GOVERNOR SPRING AND CONTROLS FOR DAMAGE OR IMPROPER INSTALLATION** → **REPLACE GOVERNOR SPRING OR CONTROLS** → **TEST RUN ENGINE**

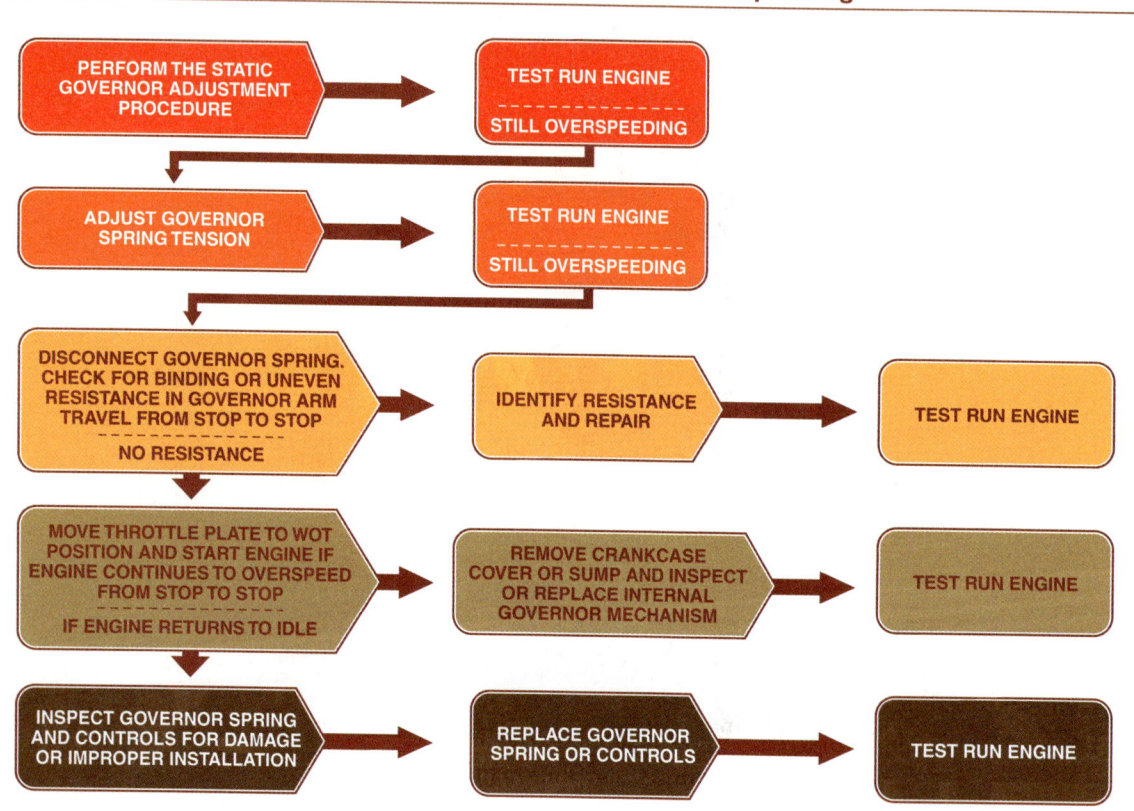

...GOVERNOR SYSTEM TROUBLESHOOTING

Pneumatic governor system - engine runs too slow

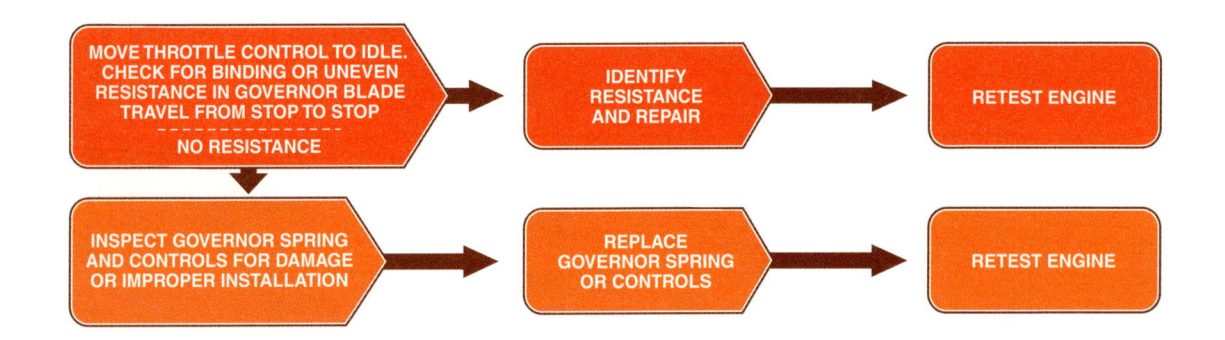

MOVE THROTTLE CONTROL TO IDLE. CHECK FOR BINDING OR UNEVEN RESISTANCE IN GOVERNOR BLADE TRAVEL FROM STOP TO STOP → IDENTIFY RESISTANCE AND REPAIR → RETEST ENGINE

NO RESISTANCE ↓

INSPECT GOVERNOR SPRING AND CONTROLS FOR DAMAGE OR IMPROPER INSTALLATION → REPLACE GOVERNOR SPRING OR CONTROLS → RETEST ENGINE

Mechanical governor system - engine runs too slow

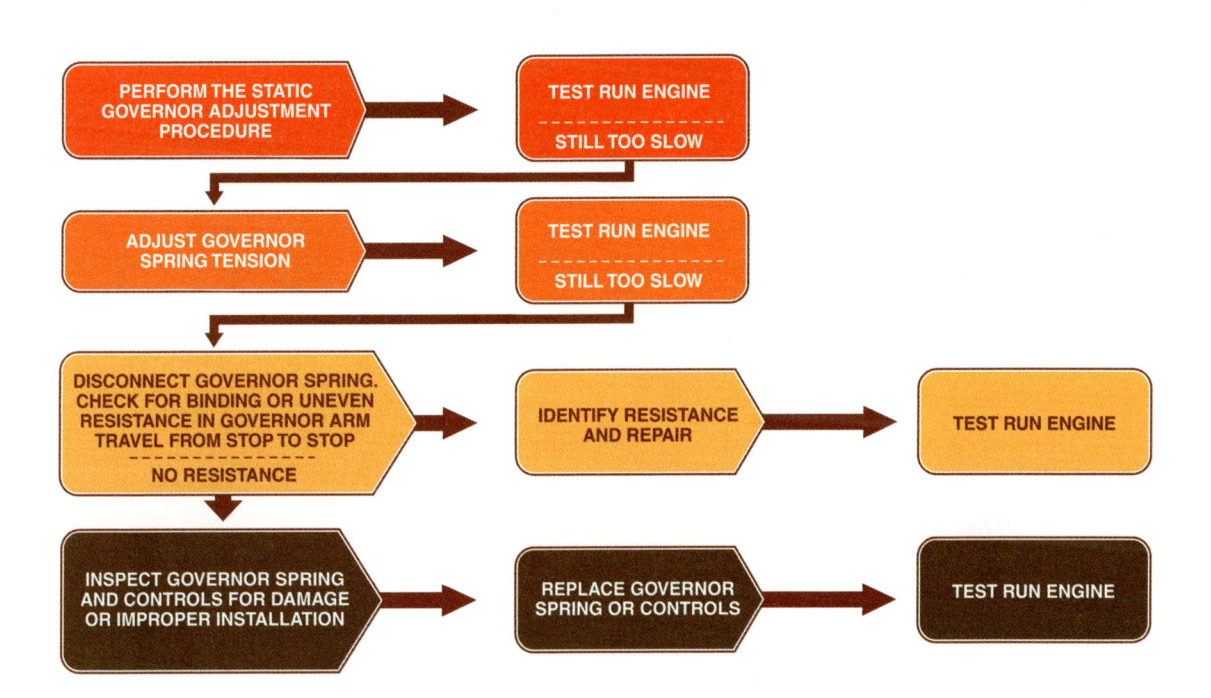

PERFORM THE STATIC GOVERNOR ADJUSTMENT PROCEDURE → TEST RUN ENGINE / STILL TOO SLOW

↓

ADJUST GOVERNOR SPRING TENSION → TEST RUN ENGINE / STILL TOO SLOW

↓

DISCONNECT GOVERNOR SPRING. CHECK FOR BINDING OR UNEVEN RESISTANCE IN GOVERNOR ARM TRAVEL FROM STOP TO STOP → IDENTIFY RESISTANCE AND REPAIR → TEST RUN ENGINE

NO RESISTANCE ↓

INSPECT GOVERNOR SPRING AND CONTROLS FOR DAMAGE OR IMPROPER INSTALLATION → REPLACE GOVERNOR SPRING OR CONTROLS → TEST RUN ENGINE

ENGLISH SYSTEM

Category		Unit	Abbreviation	Equivalents
Length		mile	mi	5280', 320 rd, 1760 yd
		rod	rd	5.50 yd, 16.5'
		yard	yd	3', 36"
		foot	ft or '	12", 0.333 yd
		inch	in. or "	0.083', 0.028 yd
Area $A = l \times w$		square mile	sq mi or mi²	640 A, 102,400 sq rd
		acre	A	4840 sq yd, 43,560 sq ft
		square rod	sq rd or rd²	30.25 sq yd, 0.00625 A
		square yard	sq yd or yd²	1296 sq in., 9 sq ft
		square foot	sq ft or ft²	144 sq in., 0.111 sq yd
		square inch	sq in. or in²	0.0069 sq ft, 0.00077 sq yd
Volume $V = l \times w \times h$		cubic yard	cu yd or yd³	27 cu ft, 46,656 cu in.
		cubic foot	cu ft or ft³	1728 cu in., 0.0370 cu yd
		cubic inch	cu in. or in³	0.00058 cu ft, 0.000021 cu yd
Capacity (WATER, FUEL, ETC.)	U.S. liquid measure	gallon	gal.	4 qt (231 cu in.)
		quart	qt	2 pt (57.75 cu in.)
		pint	pt	4 gi (28.875 cu in.)
		gill	gi	4 fl oz (7.219 cu in.)
		fluid ounce	fl oz	8 fl dr (1.805 cu in.)
		fluidram	fl dr	60 min (0.226 cu in.)
		minim	min	1/60 fl dr (0.003760 cu in.)
(VEGETABLES, GRAIN, ETC.)	U.S. dry measure	bushel	bu	4 pk (2150.42 cu in.)
		peck	pk	8 qt (537.605 cu in.)
		quart	qt	2 pt (67.201 cu in.)
		pint	pt	1/2 qt (33.600 cu in.)
(DRUGS)	British imperial liquid and dry measure	bushel	bu	4 pk (2219.36 cu in.)
		peck	pk	2 gal. (554.84 cu in.)
		gallon	gal.	4 qt (277.420 cu in.)
		quart	qt	2 pt (69.355 cu in.)
		pint	pt	4 gi (34.678 cu in.)
		gill	gi	5 fl oz (8.669 cu in.)
		fluid ounce	fl oz	8 fl dr (1.7339 cu in.)
		fluidram	fl dr	60 min (0.216734 cu in.)
		minim	min	1/60 fl dr (0.003612 cu in.)
Mass And Weight (COAL, GRAIN, ETC.)	avoirdupois	ton	t	2000 lb
		short ton	t	2000 lb
		long ton		2240 lb
		pound	lb or #	16 oz, 7000 gr
		ounce	oz	16 dr, 437.5 gr
		dram	dr	27.344 gr, .0625 oz
		grain	gr	0.037 dr, 0.002286 oz
(GOLD, SILVER, ETC.)	troy	pound	lb	12 oz, 240 dwt, 5760 gr
		ounce	oz	20 dwt, 480 gr
		pennyweight	dwt or pwt	24 gr, 0.05 oz
		grain	gr	0.042 dwt, 0.002083 oz
(DRUGS)	apothecaries'	pound	lb ap	12 oz, 5760 gr
		ounce	oz ap	8 dr ap, 480 gr
		dram	dr ap	3 s ap, 60 gr
		scruple	s ap	20 gr, 0.333 dr ap
		grain	gr	0.05 s, 0.002083 oz, 0.0166 dr ap

METRIC SYSTEM

	Unit	Abbreviation	Number of Base Units
Length	kilometer	km	1000
	hectometer	hm	100
	dekameter	dam	10
	meter*	m	1
	decimeter	dm	0.10
	centimeter	cm	0.01
	millimeter	mm	0.001
Area ($A = l \times w$)	square kilometer	sq km or km^2	1,000,000
	hectare	ha	10,000
	are	a	100
	square centimeter	sq cm or cm^2	0.0001
Volume ($V = l \times w \times h$)	cubic centimeter	cu cm, cm^3 or cc	0.000001
	cubic decimeter	dm^3	0.001
	cubic meter*	m^3	1
Capacity (WATER, FUEL, ETC.)	kiloliter	kl	1000
	hectoliter	hl	100
	dekaliter	dal	10
	liter*	l	1
	cubic decimeter	dm^3	1
	deciliter	dl	0.10
	centiliter	cl	0.01
	milliliter	ml	0.001
Mass And Weight (COAL, GRAIN, ETC. / GOLD, SILVER, ETC.)	metric ton	t	1,000,000
	kilogram*	kg	1000
	hectogram	hg	100
	dekagram	dag	10
	gram	g	1
	decigram	dg	0.10
	centigram	cg	0.01
	milligram	mg	0.001

* base units

WARRANTY CLAIM

SAFETY COLOR CODING

Color	Use	Applications
Red	Identify: 1. Fire protection equipment and apparatus 2. Danger 3. Stop	1. Fire exit signs, fire alarm boxes, fire extinguishers, fire hose locations, fire hydrants 2. Safety cans with a flash point of 100°F or less, danger signs 3. Emergency stop box on hazardous machines and stop buttons used for emergency stopping of machinery
Orange	Designate dangerous parts of machines or energized equipment which may cut, crush, shock, or otherwise cause injury and to emphasize such hazards when doors are open or guards are removed	Inside mowing guards; safety starting buttons; inside transmission guards for gears, pulleys, chains, etc.; exposed parts of pulleys, gears, rollers, etc.
Yellow	Designate caution and mark physical hazards such as striking against, stumbling, falling, and tripping. Solid yellow, yellow and black stripes, or yellow and black checks may be used in any combination to attract the most attention	Construction equipment such as bulldozers, tractors, carryalls, etc.; coverings or guards for guy wires; exposed and unguarded edges of platforms, pits, and walls; handrails, guardrails, on top and bottom treads of stairways where caution is needed; markings for projections, doorways, etc.; pillars, posts, or columns
Purple	Designate radiation hazards. Yellow is used in combination with purple for markers such as tags, labels, signs, etc.	Rooms and areas where radioactive materials are stored or handled, burial grounds, disposable cans for contaminated materials, containers of radioactive materials, etc.
Green	Designate safety and location of first aid equipment	Safety bulletin boards, gas masks, first aid kits, stretchers, etc.
Black, White, or B/W	Designate: 1. Traffic 2. Housekeeping areas	1. Dead ends of aisles or passageways; location and width of aisleways, stairways, and direction signs 2. Location of refuse cans, food dispensing equipment, etc.

EXHAUST EVACUATION SYSTEM

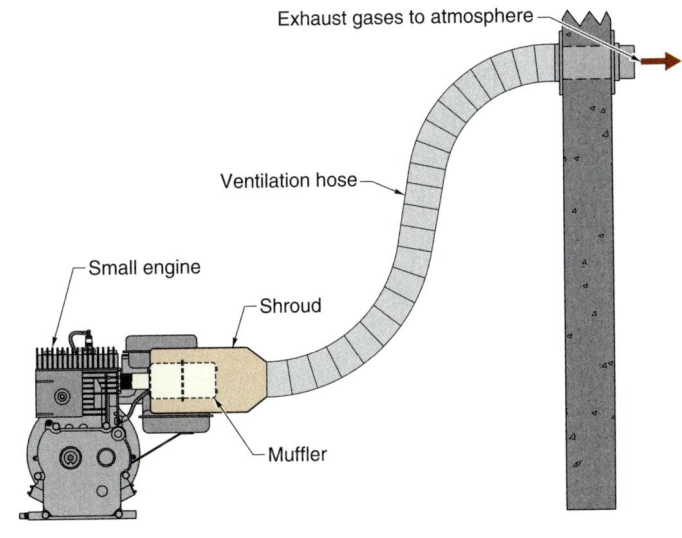

Exhaust gases to atmosphere

Ventilation hose

Small engine

Shroud

Muffler

Potential Effects of Carbon Monoxide Exposure

PPM*	Effects and Symptoms	Time
50	Permissible exposure level	8 hrs
200	Slight headache, discomfort	3 hrs
400	Headache, discomfort	2 hrs
1000-2000	Headache, discomfort	1 hr
1000-2000	Confusion, headache, nausea	2 hrs
1000-2000	Tendency to stagger	1.5 hrs
1000-2000	Slight palpitation of the heart	30 min
2000-2500	Unconciousness	30 min
4000	Fatal	< 1 hr

* values are approximate and vary with state of health and physical activity

Bacharach, Inc.

Evacuation System

HAZARDOUS MATERIAL CONTAINER LABELING...

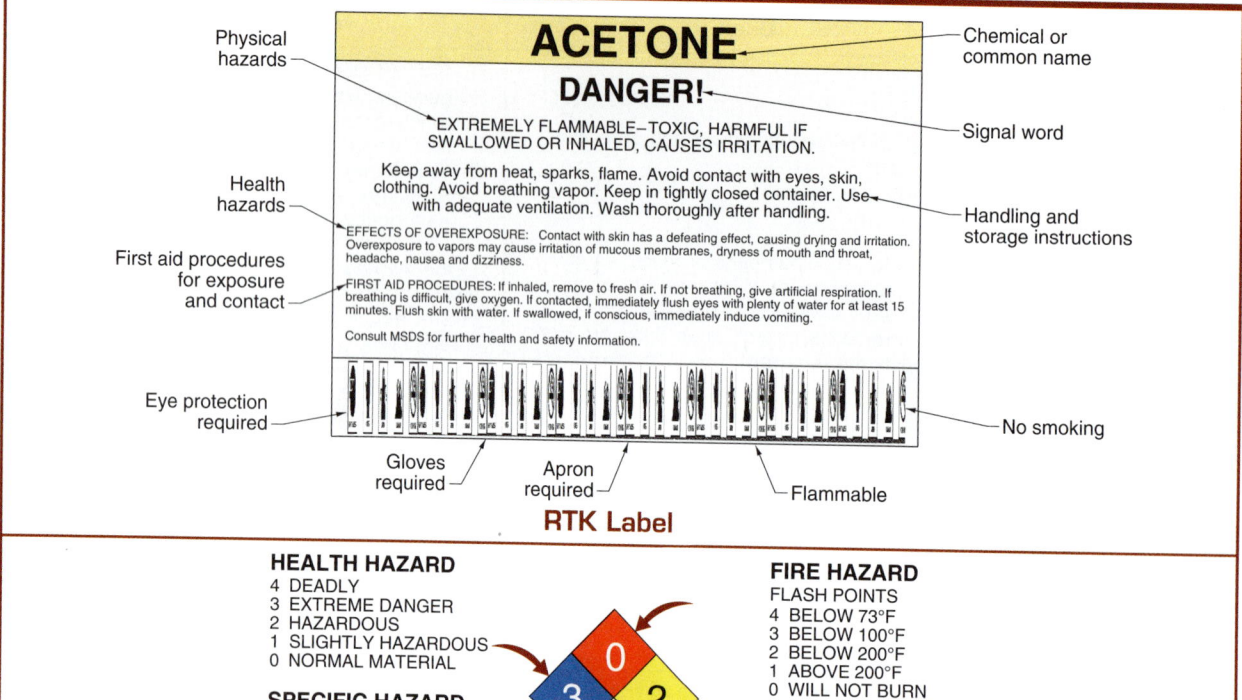

RTK Label

ACETONE — Chemical or common name

DANGER! — Signal word

EXTREMELY FLAMMABLE—TOXIC, HARMFUL IF SWALLOWED OR INHALED, CAUSES IRRITATION.

Keep away from heat, sparks, flame. Avoid contact with eyes, skin, clothing. Avoid breathing vapor. Keep in tightly closed container. Use with adequate ventilation. Wash thoroughly after handling.

EFFECTS OF OVEREXPOSURE: Contact with skin has a defeating effect, causing drying and irritation. Overexposure to vapors may cause irritation of mucous membranes, dryness of mouth and throat, headache, nausea and dizziness.

FIRST AID PROCEDURES: If inhaled, remove to fresh air. If not breathing, give artificial respiration. If breathing is difficult, give oxygen. If contacted, immediately flush eyes with plenty of water for at least 15 minutes. Flush skin with water. If swallowed, if conscious, immediately induce vomiting.

Consult MSDS for further health and safety information.

Callouts: Physical hazards; Health hazards; First aid procedures for exposure and contact; Eye protection required; Gloves required; Apron required; Handling and storage instructions; No smoking; Flammable.

NFPA Hazard Signal System

HEALTH HAZARD
- 4 DEADLY
- 3 EXTREME DANGER
- 2 HAZARDOUS
- 1 SLIGHTLY HAZARDOUS
- 0 NORMAL MATERIAL

SPECIFIC HAZARD
- OX   OXIDIZER
- ACID ACID
- ALK   ALKALI
- COR   CORROSIVE
- W̶   USE NO WATER
- ☢   RADIATION HAZARD

FIRE HAZARD — FLASH POINTS
- 4 BELOW 73°F
- 3 BELOW 100°F
- 2 BELOW 200°F
- 1 ABOVE 200°F
- 0 WILL NOT BURN

REACTIVITY
- 4 MAY DETONATE
- 3 SHOCK AND HEAT MAY DETONATE
- 2 VIOLENT CHEMICAL CHANGE
- 1 UNSTABLE IF HEATED
- 0 STABLE

Diamond values shown: Health 3 (blue), Fire 0 (red), Reactivity 2 (yellow), Specific W̶ (white).

Identification of Health Hazard — Color Code: BLUE		Identification of Flammability — Color Code: RED		Identification of Reactivity (Stability) — Color Code: YELLOW	
Signal	Type of Possible Injury	Signal	Susceptibility of Materials to Burning	Signal	Susceptibility to Release of Energy
4	Materials that on very short exposure could cause death or major residual injury	4	Materials that will rapidly or completely vaporize at atmospheric pressure and normal ambient temperature, or that are readily dispersed in air and that will burn readily	4	Materials that in themselves are readily capable of detonation or reaction at normal temperatures and pressures
3	Materials that on short exposure could cause serious temporary or residual injury	3	Liquids and solids that can be ignited under almost all ambient temperature conditions	3	Materials that in themselves are capable of detonation or explosive decomposition or reaction but require a strong initiating source or which must be heated under confinement before initiation or which react explosively with water
2	Materials that on intense or continued but not chronic exposure could cause temporary incapacitation or possible residual injury	2	Materials that must be moderately heated or exposed to relatively high ambient temperatures before ignition can occur	2	Materials that readily undergo violent chemical change at elevated temperatures and pressures or which react violently with water or which may form explosive mixtures with water
1	Materials that on exposure would cause irritation but only minor residual injury	1	Materials that must be preheated before ignition can occur	1	Materials that in themselves are normally stable, but which can become unstable at elevated temperatures and pressures
0	Materials that on exposure under fire conditions would offer no hazard beyond that of ordinary combustible material	0	Materials that will not burn	0	Materials that in themselves are normally stable, even under fire exposure conditions, and which are not reactive with water.

Reprinted with permission from NFPA704-1990, *Identification of the Fire Hazards of Materials*, Copyright ©1990, National Fire Protection Association, Quincy, MA 02269. This reprinted material is not the complete and official position of the National Fire Protection Association, on the referenced subject which is represented only by the standard in its entirety.

...HAZARDOUS MATERIAL CONTAINER LABELING

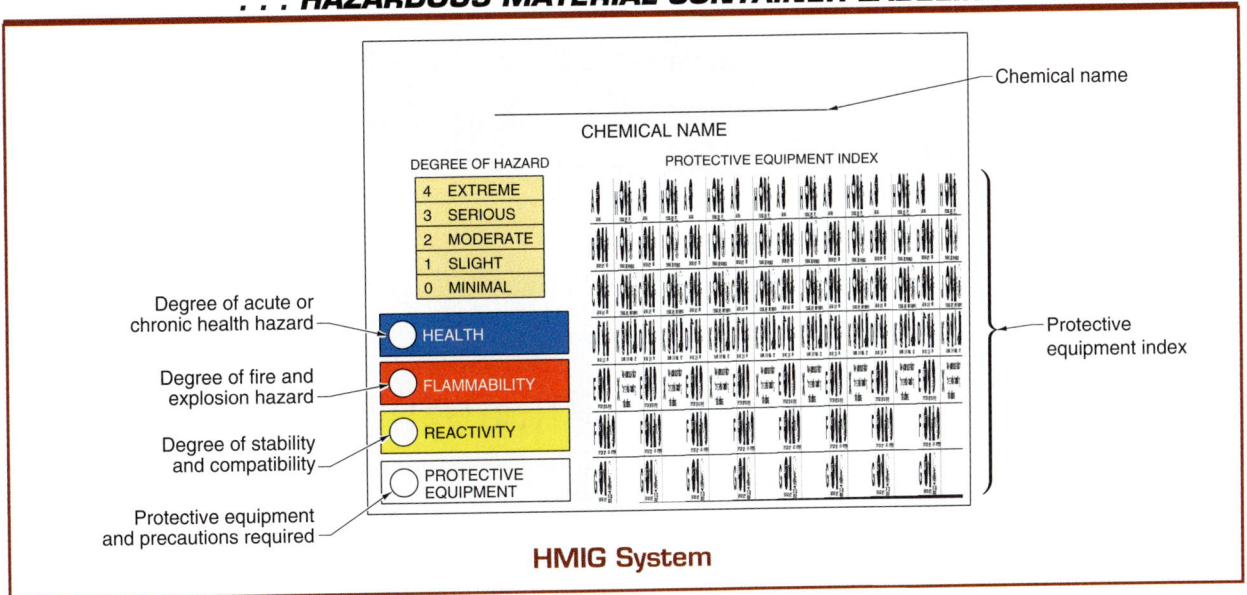

HMIG System

- Chemical name
- Protective equipment index

DEGREE OF HAZARD
- 4 EXTREME
- 3 SERIOUS
- 2 MODERATE
- 1 SLIGHT
- 0 MINIMAL

- Degree of acute or chronic health hazard — HEALTH
- Degree of fire and explosion hazard — FLAMMABILITY
- Degree of stability and compatibility — REACTIVITY
- Protective equipment and precautions required — PROTECTIVE EQUIPMENT

HAZARDOUS MATERIALS

Chemical	H	F	R	S/H	Chemical Abstract Service Number	Chemical	H	F	R	S/H	Chemical Abstract Service Number
Acetic acid	2	2	0	—	64-19-7	Isopropyl ether	2	3	1	—	108-20-3
Acetone	1	3	0	—	67-64-1	Methanol	1	3	0	—	67-56-1
Acetonitrile	3	3	0	—	75-05-8	Methyl acetate	1	3	0	—	79-20-9
Acrolein	3	3	3	—	107-02-8	Methyl bromide	3	1	0	—	74-83-9
Allyl alcohol	3	3	1	—	107-16-6	Methyl isobutyl ketone	2	3	0	—	108-10-1
Ammonia anhydrous	3	1	0	—	7664-41-7	Methylamine	3	4	0	—	74-89-5
Aniline	3	2	0	—	65-53-3	Morpholine	2	3	0	—	110-91-8
Bromine	3	0	0	OX	7726-95-6	Naphtha	1	4	0	—	8030-30-6
1 – 3 Butadiene	2	4	2	—	106-99-0	Naphthalene	2	2	0	—	91-20-3
Butyl acetate	1	3	0	—	123-86-4	Nitric acid	3	0	0	OX	7697-37-2
tert-Butyl alcohol	1	3	0	—	75-65-0	Nitrobenzene	3	2	1	—	98-95-3
Caustic soda	3	0	1	—	1310-73-2	p-Nitrochlorobenzene	2	1	3	—	100-00-5
Chlorine	3	0	0	OX	7782-50-5	Octane	0	3	0	—	111-65-9
Chloroform	2	0	0	—	67-66-3	Oxalic acid	2	1	0	—	144-62-7
o-Cresol	3	2	0	—	1319-77-3	Pentane	1	4	0	—	109-66-0
Cumene	2	3	1	—	98-82-8	Petroleum distillates	1	4	0	—	8002-05-9
Cyclohexane	1	3	0	—	110-82-7	Phenol	3	2	0	—	108-95-2
Cyclohexanol	1	2	0	—	108-93-0	Propane gas	1	4	0	—	74-98-6
Cyclohexanone	1	2	0	—	108-94-1	1-Propanol	1	3	0	—	71-23-8
Diborane	3	4	3	W	19287-45-7	Propyl acetate	1	3	0	—	109-60-7
Dimethylamine	3	4	0	—	124-40-3	n-Propyl alcohol	1	3	0	—	71-23-8
p-Dioxane	2	3	1	—	123-91-1	Propylene oxide	4	2	2	—	75-56-9
Ethyl acetate	1	3	0	—	141-78-6	Pyridine	2	3	0	—	110-86-1
Ethyl ether	2	4	1	—	60-29-7	Sodium cyanide	3	0	0	—	143-33-9
Formic acid	3	2	0	—	64-18-6	Sodium hydroxide	3	0	1	—	1310-73-2
n-Heptane	1	3	0	—	142-82-5	Stoddard solvent	0	2	0	—	8052-41-3
n-Hexane	1	3	0	—	110-54-3	Sulfur dioxide	3	0	0	—	7446-09-5
Hydrazine	3	3	3	—	302-01-2	Sulfuric acid	3	0	2	W	7664-93-9
Hydrochloric acid	3	0	0	—	7647-01-0	Tetrahydrofuran	2	3	1	—	109-99-9
Hydrogen peroxide	2	0	1	OX	7722-84-1	1-1-1 Trichloroethane	2	1	0	—	71-55-6
Iodine	—	—	—	—	7553-56-2	Triethylamine	2	3	0	—	121-44-8
Isobutyl alcohol	1	3	0	—	78-83-1	Xylene	2	3	0	—	1330-20-7
Isopropyl alcohol	1	3	0	—	67-63-0						

Lab Safety Supply, Inc.

194 SMALL ENGINE AND EQUIPMENT MAINTENANCE

HORSEPOWER TO TORQUE CONVERSION

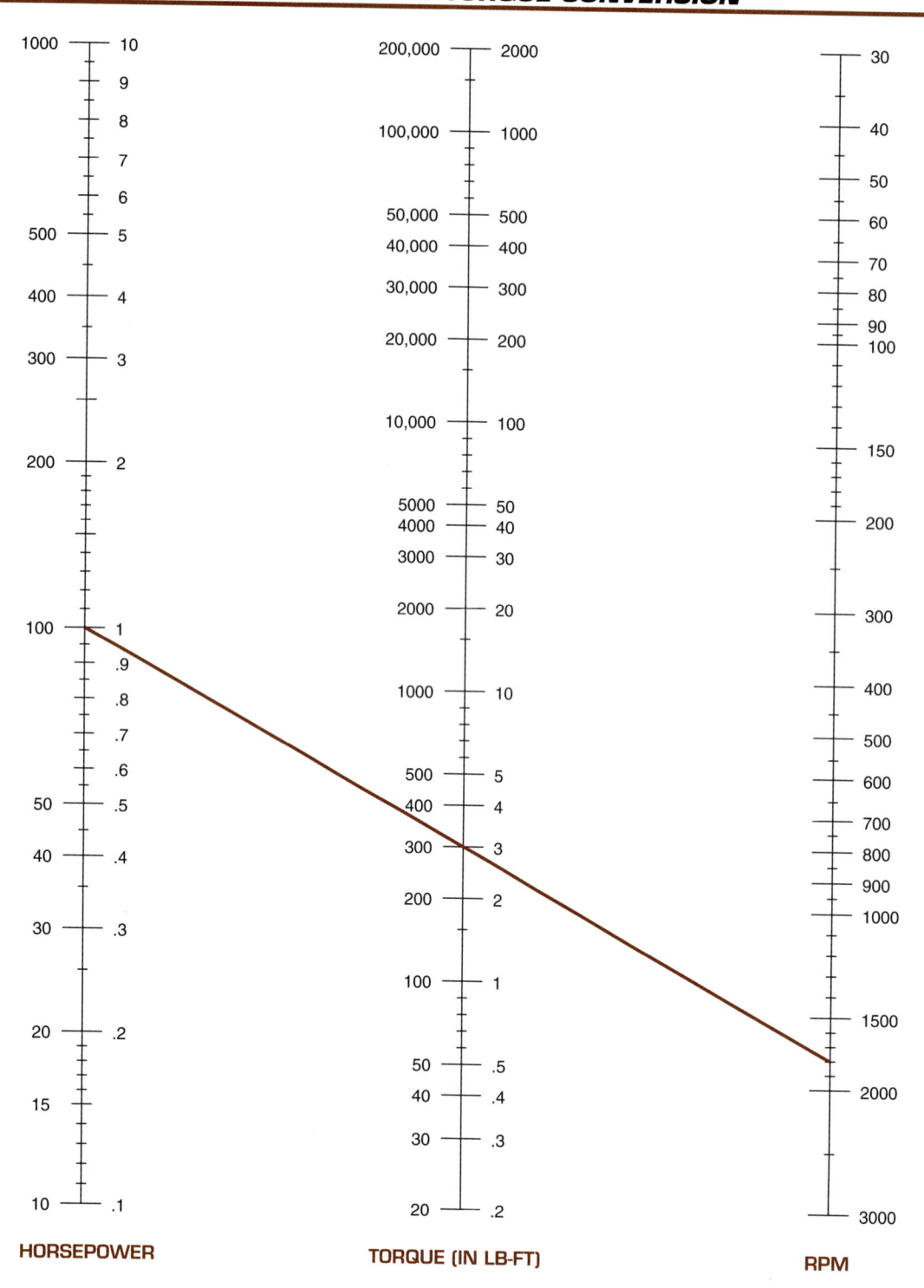

HORSEPOWER	TORQUE (IN LB-FT)	RPM

Glossary

A

air cleaner: A device designed to remove airborne impurities, such as dust, fumes, gases, and vapors, from the air.

air vane: A device that monitors engine RPM so that a governor can maintain the selected engine speed. Also known as a flyweight.

alternating current (AC): Electrical current flow that reverses direction at regular intervals.

alternator: A charging system device that produces alternating current.

armature: The rotating part of a generator that consists of a segmented iron core surrounded by copper wires wound tightly together.

attachment: A device connected to a machine or implement that is used to perform a task.

B

belt guide: A component used with a belt drive system to retain a flexible belt within the confines of a pulley groove.

blower housing: A sheet metal or composite, such as plastic, component that encompasses a fan to direct cooling air to a cylinder block and cylinder head.

breaker point: An ignition system component that has two points (contact surfaces) that together function as a mechanical switch.

C

carbon monoxide (CO): An odorless, colorless, and poisonous gas produced by burning gasoline and other fuels.

carburetor: A device that draws in fuel from a fuel tank and outside air to form a combustible vapor that is fed into the combustion chamber of an engine.

choke plate: A flat plate placed in a carburetor body between the throttle plate and air intake that is used to restrict airflow to help start a cold engine.

combustion: A rapid, oxidizing chemical reaction in which a fuel chemically combines with atmospheric oxygen and releases energy in the form of heat.

combustion event: *See* ignition event.

commercial and industrial outdoor power equipment: Outdoor power equipment that is used exclusively outdoors in a commercial or industrial environment.

compression event: An engine operation event in which a trapped air-fuel mixture is compressed inside a combustion chamber.

compression ring: A piston ring located in the ring groove closest to a piston head.

concrete mixer: Portable or stationary equipment that consists of a rotating drum or paddle used to mix concrete ingredients.

concrete vibrator: A power tool used to agitate and consolidate freshly placed concrete and produce close contact with a form or mold.

cooling fin: An integral thin cast strip designed to provide efficient air circulation and the dissipation of heat away from an engine cylinder block into the air stream.

crankcase: An engine component that houses and supports a crankshaft.

crankshaft: An engine component, such as a lawn mower blade or snow thrower auger, that converts the linear (reciprocating) motion of a piston and connecting rod into rotary motion.

cylinder head: A cast aluminum alloy or cast iron engine component fastened to the end of the cylinder block farthest from a crankshaft.

D

direct current (DC): Electrical current that flows in one direction only.

disc drive: A power transmission device that uses a friction disc to make contact between the driving and driven components of equipment.

E

engine block: The main structure of an engine that supports and helps maintain the alignment of internal and external components.

exhaust event: An engine operation event in which spent gases are removed from a combustion chamber and released into the atmosphere.

F

flame front: A boundary wall that separates the charge from combustion by-products.

flexible belt drive: A mechanical drive system that uses a flexible belt to transfer power between a drive and driven shaft.

flyweight: *See* air vane.

flywheel: A cast iron, aluminum, or zinc disk that is mounted on one end of a crankshaft to provide inertia for an engine to prevent the loss of engine speed between combustion intervals.

flywheel brake: A device that is included on the engine of equipment that requires constant operator presence such as lawn mowers.

four-stroke cycle engine: An internal combustion engine that uses four piston strokes (intake, compression, power, and exhaust) to complete one operating cycle.

fuel stabilizer: A compound used to extend the life of fuel that is not or cannot be stored properly.

fuel tank: A liquid storage vessel connected to an engine for the purpose of holding fuel, such as gasoline or diesel fuel, to power the engine.

G

governor system: An engine system that maintains a desired engine speed regardless of the load applied to the engine.

H

hitch pin: A pin used to secure a locator dowel when connecting two pieces of equipment.

hydrostatic drive: A power transmission device that uses pressurized fluid to provide power to equipment without direct contact between driving and driven components.

I

ignition event: An engine operation event in which a charge is ignited and rapidly oxidized through a chemical reaction to release heat energy. Also known as a combustion event.

ignition switch: A rotary-actuated, double-pole double-throw (DPDT) switch that uses a key as an actuating mechanism.

implement: A device connected to, and typically powered by, a machine that is used to perform a specific task.

intake event: An engine operation event in which an air-fuel mixture is introduced to a combustion chamber.

L

lawn and garden power equipment: Equipment that is used exclusively outdoors in residential or light commercial environments.

lawn and garden tractor: A gasoline-powered four wheel machine that allows the operator to ride on the equipment as it performs work.

leaf blower: A motorized device that creates high-speed and high-volume blowing air to move light debris away from driveways, walkways, roofs, and lawns.

locator dowel: A cylindrical metal rod inserted into the holes in the adjacent members of a joint to align and strengthen the joint.

M

mechanical drive system: A drive system used to transfer power from one location to another, typically from an engine to an implement or equipment drive train.

mechanical governor: An engine governor that adjusts the throttle plate position as needed by using the gears and flyweights inside a crankcase as speed-sensing devices to detect changes in a load.

mechanical governor system: An engine system consisting of a gear assembly that meshes with a camshaft or other engine components to sense and maintain a desired engine speed.

mechanical switch: An electrical control device used to allow, interrupt, or direct the flow of electricity through a circuit.

mower deck: *See* rotary mower deck.

muffler: An engine component fitted with baffles and plates that subdues noise produced from exhaust gases exiting a combustion chamber.

O

operator presence control switch: A switch used to ensure the safe positioning of an equipment operator relative to the equipment.

P

piston: A cylindrical engine component that slides back and forth in a cylinder bore by forces produced during a combustion process.

piston ring: An expandable split ring used to provide a seal between a piston and cylinder bore.

pneumatic governor: An engine governor that uses a movable metal or plastic air vane as a speed-sensing device to detect changes in a load.

portable generator: A portable machine that changes rotating mechanical energy into electric energy.

power event: An engine operation event in which a compressed charge is ignited and hot rapidly expanding gases force a piston head away from a cylinder head.

power take-off (PTO) switch: A switch used to control the PTO function on a lawn and garden tractor.

pressure washer: A portable device that cleans surfaces with pressurized water.

primer bulb: An enrichment system consisting of a rubber bulb filled with fuel or air connected to a fuel bowl by a passageway.

pulley: A grooved rotating wheel with a belt around a portion of its circumference that is used for transferring power.

push reel lawn mower: A lawn cutting device with multiple cutting blades attached to a rotating central shaft between two wheels.

R

rewind cord: A device that is pulled to start the combustion process within an engine ("start the engine").

rotary lawn mower: A grass cutting device that uses one or more flat, horizontal blades to cut grass with a high-speed circular rotation.

rotary mower deck: An implement that uses one or more flat blades to cut grass with a circular motion. Also known as a mower deck.

S

safety interlock switch: An electrical control device that helps prevent the unsafe operation of equipment.

shroud: A rigid, protective cover attached to an engine near the fuel tank to protect the engine from moisture, dust, dirt, and other debris.

snow thrower: A machine that removes snow, as from a sidewalk or driveway, by using a rotating spiral blade to pick up the snow and propel it aside.

soil compactor: A mechanical or manual device used to improve the bearing capacity of soil by tamping or vibrating the soil.

spark arrestor: A component in the exhaust system of a small engine that redirects the flow of exhaust gases through a screen to trap sparks discharged from the engine.

spark gap: The distance from the center electrode to the ground electrode of a spark plug.

spark plug: A component that isolates electricity induced in secondary windings and directs a high-voltage charge to the spark gap at the tip of a spark plug.

spark tester: A test tool used to test the ignition system of a small engine.

stator: An electrical component that has a continuous copper wire (stator winding) wound on separate stubs exposing the wire to a magnetic field.

string trimmer: A powered handheld device that uses a flexible monofilament line for cutting grass and other plants near objects.

T

throat: The main passageway of a carburetor that directs an air-fuel mixture and air from the atmosphere to a combustion chamber.

throttle plate: A disk that pivots on a movable shaft to regulate the air and fuel flow in a carburetor.

V

valve: An engine component that opens and closes at precise times to allow the flow of an air-fuel mixture into and exhaust gases from a cylinder.

V belt: A reinforced rubberized belt that connects pulleys to rotate components.

venturi: The narrow portion of a carburetor throat.

vibratory plate: Gasoline- or diesel-powered compaction equipment used to compact the surface of pavement and/or pavement underlayment materials.

W

wiper ring: A piston ring with a tapered face located in the ring groove between the compression ring and oil ring.

wood chipper: A device used to chop and shred organic materials into ¼″ particles.

Z

zero-turn lawn mower: A ride-on lawn mower with a tight turning radius that allows the operator to control forward and backward motion using control handles.

Index

Page numbers in italic refer to figures.

A

AC, *60*, 60–61
accessory switches, 69
AC voltage testing, *150*, 150
adjustable pliers, 1, *2*
adjustable wrenches, 1, *2*
adjusting air gaps on external stators, *154*, 154
adjusting carburetors, 124, *125*
adjusting choke linkages, *129*, 129
adjusting governed idle speed, 134
adjusting idle speed screws, 126, *127*
adjusting mixture screws, 126–128, *127*, *128*
adjusting static settings on mechanical governors, 132–133, *133*
adjusting tappet clearances, 169
adjustment screws, 134, *135*
air cleaners, 44, *45*, *46*
air cleaners, servicing, 90–93, *91*, *92*, *93*
air vanes, 44
Allen wrenches, *2*, 2
alternating current (AC), *60*, 60–61
alternators, 59, *60*
alternator tests, 148, *149*
aluminum engines, 18, *21*
armatures, 33
attachments, 69, *70*
augers, 72

B

basters, 1, *2*
battery safety, 148
battery storage, *79*, 79
belt drives, 64–66, *65*, *66*, 73–74, *74*
belt guides, 66
blades, mower, *71*, 71–72
blower fans, 72
blower housings, 44, *45*, *46*, *47*
brake bands, testing, *117*, 117–118, *118*
brake pad systems, testing, 114–116, *115*, *116*
braking systems, 61, *62*
breaker point ignition systems, 54, *55*
breaker points, 54, *55*

C

carbon deposit removal, 142–146, *143*, *145*, *147*
carbon monoxide (CO) meters, *6*, 6
carburetor overhauling, *106*, 106–111, *107*, *109*, *111*
carburetors, 44, *45*, *46*, 52–53, *53*
carburetors, adjusting, 124, *125*
carburetors, cleaning, 124, *125*
cast iron engines, 14–16, *15*
center punches, 1, *2*
changing engine oil, *82*, 82–83, *83*
choke linkages, adjusting, *129*, 129
choke plates, *53*, 53, *54*
cleaning carburetors, 124, *125*
cleaning fuel tanks, *119*, 119
cleaning OPE, 72, *73*
clothes washing machines, *17*, 17
combination wrench sets, 1, *2*
combustion, 40
combustion event, 40, *41*, *42*, *43*
commercial OPE, 32–33, *33*
compaction equipment, 34–35, *35*
components, 44–47, *45*, *46*, *47*
compression event, 40, *40*, *43*
compression rings, *50*, 50
compression systems, *48*, 48–50, *49*, *50*, *51*
concrete mixers, *34*, 34
concrete vibrators, 35
consumer OPE, *29*, 29–32, *30*, *31*, *32*
cooling fins, *45*, 45
cooling systems, *56*, 56
crankcases, *45*, 45, *47*, *50*, *51*
crankshafts, *45*, 45, *46*, *47*
cylinder heads, *45*, 45, *46*, *47*
cylinder heads, reassembling, 146–147, *147*

D

DC, *60*, 60–61
DC current testing, *151*, 151
debris removal, 84–86, *85*, *86*
diagonal wire cutters, 1, *2*
dipsticks, 44, *45*, *46*, *47*
direct current (DC), *60*, 60–61
direct overhead valve (DOV) engines, *48*, 48
disc drives, 64, *65*, 73–74, *75*
DOV engines, *48*, 48
dowels, 76
drills, *3*, 3
drive belts, replacing, 96, *97*
drive discs, replacing, 140, *141*
drive system maintenance, 73–74, *74*, *75*

E

edgers, 20
electrical systems, 59–61, *60*
electrical system service, 148–154
 adjusting air gaps on external stators, *154*, 154
 alternator tests, 148, *149*
 battery safety, 148
 replacing external stators, *153*, 153
 replacing stators under flywheels, *152*, 152
 testing AC voltage, *150*, 150
 testing DC current, *151*, 151
engine blocks, 44, *45*, *46*
engine bolts, tightening, 101
engine components, 44–47, *45*, *46*, *47*
engine fuel safety, *8*, 8–9, *9*
engine general maintenance, *100*, 100–101, *101*
engine maintenance, *73*, 73
engine oil, 44
engine oil, changing, *82*, 82–83, *83*
engine speed-control levers, *130*, 131
engine storage, 77, *78*
engine troubleshooting, *100*, 100–101, *101*
equipment storage, 78–79, *79*
erratic engine behavior, troubleshooting, 132
exhaust event, *42*, 42, *43*
exhaust valves, 44, *45*, *46*
external stators, *153*, 153–154, *154*

F

feeler gauges, *4*, 4
flame fronts, 40
flat files, 1, *2*
flathead screwdrivers, 1, *2*
flexible belt drives, 64–66, *65*, *66*, 73–74, *74*
flyweights, *130*, 131
flywheel brakes, 44, *45*, *46*
flywheel brake service, 114–118, *115*, *116*, *117*, *118*
flywheel holders, *4*, 4
flywheel key replacement, 112, *113*
flywheel magnets, 44, *45*
flywheel pullers, *4*, 4
flywheel replacement, 112, *113*
flywheels, 44, *45*, *46*, *47*
four-stroke cycle engines, 23, *24*, 37–43, *37–42*

199

200 SMALL ENGINE AND EQUIPMENT MAINTENANCE

fuel filters, inspecting, *120,* 120–121, *121*
fuel line clamp tools, *4,* 4
fuel pumps, servicing, 122, *123*
fuel stabilizers, 77, *78*
fuel systems, 51–53, *52, 53, 54*
fuel system service, 119–129
    adjusting carburetors, 124, *125*
    adjusting choke linkages, *129,* 129
    adjusting idle speed screws, 126, *127*
    adjusting mixture screws, 126–128, *127, 128*
    cleaning carburetors, 124, *125*
    cleaning fuel tanks, *119,* 119
    gasoline usage, 121
    inspecting fuel filters, *120,* 120–121, *121*
    servicing fuel pumps, 122, *123*
fuel tanks, 44, *45, 46,* 47
fuel tanks, cleaning, *119,* 119
fuel safety, *8,* 8–9, *9*

G

gasoline usage, 121
general maintenance, small engines, *100,* 100–101, *101*
general-purpose tools, 1–3, *2–3*
generators, portable, *20, 33,* 33
governed idle speed, adjusting, 134
governor control linkages, *130,* 131
governor components, *130,* 130–131
governor systems, *57,* 57–59, *58, 59*
governor system service, 130–137
    adjusting governed idle speed, 134
    adjusting static settings, 132–133, *133*
    mechanical governor systems, *130,* 130–131, *131*
    setting top no-load speeds, *136,* 136–137, *137*
    tang bending, 134, *135*
    troubleshooting hunting and surging, 132
governor tang benders, 4–5, *5*

H

hacksaws, 1, *2*
hex wrench sets, *2,* 2
high-speed mixture screws, adjusting, *128,* 128
hitch pins, 76, *77*
hunting and surging, troubleshooting, 132
hydrostatic drives, 63, *64,* 73

I

idle mixture screws, adjusting, 126, *127*
idle speed screws, adjusting, 126, *127*
ignition armatures, *155,* 155–156, *157,* 160, *161*
ignition event, 40, *41, 42, 43*
ignition switches, *67,* 68
ignition systems, 54–55, *55*
ignition systems, replacing, *155,* 155–160, *157, 159, 161*
    replacing ignition armatures, 156, *157*
    retrofitting old ignition armatures, 160, *161*

testing stop switches, 158, *159*
ignition systems, servicing, *87,* 87–89, *88, 89*
implement maintenance, 74–77, *76, 77*
implements, 69–72, *70, 71, 72*
industrial OPE, 32–35, *33, 34, 35*
inspecting fuel filters, *120,* 120–121, *121*
intake event, *40,* 40
intake valves, 44, *45,* 46

L

lapping valves, *168,* 168
lawn and garden power equipment, 24–29, *25–29*
lawn and garden tractors, *22,* 22, 28, *29*
lawn edgers, 20
lawn mower blades, *71,* 71–72
lawn mowers
    push reel, *18,* 18
    rotary, *21,* 21, 24–27, *25–27*
    zero-turn, *27,* 27–28, *28*
leaf blowers, *19, 30,* 30
L-head valving systems, *48,* 48
locator dowels, 76
locking pliers, 2, *3*
long-term storage, *102,* 102–104, *103, 104*
lubrication systems, *56,* 56

M

machining valves, 164
maintenance and repair safety, 9
market growth, consumer equipment, 18–22, *19, 20, 21, 22*
measuring rulers, 2, *3*
mechanical drive systems, 63–66, *64, 65, 66*
mechanical governors, *58,* 58
mechanical governor static settings, adjusting, 132–133, *133*
mechanical governor systems, *130,* 130–131, *131*
mechanical switches, *67,* 67–68, *69*
mixture screws, adjusting, 126–128, *127, 128*
Model P engines, 14, *15,* 16
mower blades, *71,* 71–72
mower-deck drive belts, replacing, 96, *97*
mowers
    push reel, *18,* 18
    rotary, *21,* 21, 24–27, *25–27*
    zero-turn, *27,* 27–28, *28*
mufflers, 44, *45, 46*
mufflers, removing, *94,* 94–95, *95*
multimeters, *4,* 4

N

needle-nose pliers, 2, *3*

O

OHVs, *43,* 43, *48,* 48
oil evacuator pumps, *4,* 4

OPE. *See* outdoor power equipment (OPE)
OPE maintenance, 72–77, *73–77*
operating condition safety, 6–7, *7*
operating cycle, four-stoke cycle engines, 38–42, *39, 40, 41, 42*
operating cycle, two-stroke cycle engines, 42, *43*
operator presence control switches, *67,* 68
operator safety, *11,* 11
operator's manuals, 79
OPE storage, 77–80, *78, 79, 80*
outdoor power equipment (OPE)
    commercial OPE, 32–33, *33*
    consumer market growth, 18–22, *19, 20, 21, 22*
    consumer OPE, *29,* 29–32, *30, 31, 32*
    early consumer equipment, *16,* 16–18, *17, 18*
    early small engines, 13–16, *14, 15*
    four-stroke cycle engine applications, 23, *24*
    industrial OPE, 32–35, *33, 34, 35*
    lawn and garden power equipment, 24–29, *25–29*
    maintenance, 72–77, *73–77*
    storage, 77–80, *78, 79, 80*
overhauling carburetors, *106,* 106–111, *107, 109, 111*
overhead valves (OHVs), *43,* 43, *48,* 48
owner's manuals, 79

P

parts cleaning brushes, 2, *3*
personal protective equipment (PPE), *11,* 11
Phillips-head screwdrivers, 2, *3*
piston rings, 49–50, *50*
pistons, 44, *45–47, 49,* 49
pliers, 1, *2, 3,* 3
pneumatic governors, 58–59, *59*
portable generators, *20, 33,* 33
power drills, *3,* 3
power event, *41,* 41, 42, *43*
power takeoff (PTO) switches, 68, *69*
PPE, *11,* 11
pressure washers, *19, 29,* 29
primer bulbs, 44, *45,* 46
PTO switches, 68, *69*
pulleys, *66,* 66
push reel lawn mowers, *18,* 18
putty knives, *3,* 3

R

ratchet torque wrenches, *4,* 4
ratchet wrenches and socket sets, *3,* 3
recoil system service, *138,* 138
removing carbon deposits, 142–146, *143, 145, 147*
removing debris, 84–86, *85, 86*
removing mufflers, *94,* 94–95, *95*
replacing drive belts, 96, *97*
replacing drive discs, 140, *141*

Index

replacing external stators, *153*, 153
replacing flywheel keys, 112, *113*
replacing flywheels, 112, *113*
replacing ignition systems, *155*, 155–160, *157*, *159*, *161*
replacing mower-deck drive belts, 96, *97*
replacing shear pins, *98*, 98
replacing skid shoes, *99*, 99
replacing stators, *152*, 152–153, *153*
replacing valve seat inserts, 168
rewind cords, 44, *45*
rewind starter system service, *138*, 138
rotary lawn mowers, *21*, 21, 24–27, *25–27*
rotary mower deck maintenance, 75–76, *76*
rotary mower decks, *71*, 71–72
rototillers, 20

S

safety, 6–12
 battery safety, 148
 engine fuel safety, *8*, 8–9, *9*
 maintenance and repair safety, 9
 operating conditions, 6–7, *7*
 operator safety, *11*, 11
 spark plug safety, *12*, 12
 tool safety, 10
safety interlock switches, *67*, 67–68
screwdrivers, 1, *2*, *3*
seasonal storage, *80*, 80
servicing air cleaners, 90–93, *91*, *92*, *93*
servicing electrical systems. *See* electrical system service
servicing flywheel brakes, 114–118, *115*, *116*, *117*, *118*
servicing fuel pumps, 122, *123*
servicing fuel systems. *See* fuel system service
servicing governor systems. *See* governor system service
servicing ignition systems, *87*, 87–89, *88*, *89*
servicing recoil systems, *138*, 138
servicing rewind starter systems, *138*, 138
servicing valves. *See* valve service
setting top no-load speeds, *136*, 136–137, *137*
shear pins, *98*, 98
shot-filled mallets, *3*, 3
shrouds, 44, *45*, *46*, *47*
single-stage snow throwers, 72
skid shoes, *99*, 99
slip-joint pliers, *3*, 3
small engine components, 44–47, *45*, *46*, *47*
small engine general maintenance, *100*, 100–101, *101*
small engine maintenance, *73*, 73
small engine storage, 77, *78*
small engine systems, 48–62
 braking systems, 61, *62*
 compression systems, *48*, 48–50, *49*, *50*, *51*
 cooling systems, *56*, 56
 electrical systems, 59–61, *60*
 fuel systems, 51–53, *52*, *53*, *54*
 governor systems, *57*, 57–59, *58*, *59*
 ignition systems, 54–55, *55*
 lubrication systems, *56*, 56
small engine tools, *4*, 4–5, *5*
small engine troubleshooting, *100*, 100–101, *101*
snow thrower maintenance, 76–77, *77*
snow throwers, *19*, 31–32, *32*, *72*, 72
snow thrower shear pins, replacing, *98*, 98
snow thrower skid shoes, replacing, *99*, 99
soil compactors, *35*, 35
solid-state ignition systems, *55*, 55
spark arrestors, 6, *7*
spark gaps, 45
spark plug gauges, *4*, *5*, *87*
spark plug leads, 45, *47*
spark plugs, *45*, 45, *46*
spark plug safety, *12*, 12
spark testers, *4*, *5*, *88*, 88–89, *89*
standard pliers, *3*, 3
star-shaped (Torx®) driver sets, *3*, 3
starter clutch wrenches, *4*, *5*
stators, 59, *60*
stators, replacing, *152*, 152–153, *153*
stop switches, checking, 84
stop switches, testing, 158, *159*
string trimmers, *19*, *31*, 31
switches, *67*, 67–68, *69*

T

tachometers, *4*, *5*
tang benders, 4–5, *5*
tang bending, 134, *135*
tappet clearances, *163*, 163, 169
testing AC voltage, *150*, 150
testing brake bands, *117*, 117–118, *118*
testing brake pad systems, 114–116, *115*, *116*
testing DC current, *151*, 151
testing stop switches, 158, *159*
throats, 52, *53*
throttle levers, *130*, 131
throttle linkages, *130*, 131
throttle plates, *53*, 53, *54*
tightening engine bolts, 101
tools, 1–5, *2–5*
tool safety, 10
top no-load speeds, setting, *136*, 136–137, *137*
Torx® driver sets, *3*, 3
troubleshooting erratic engine behavior, 132
troubleshooting hunting and surging, 132
troubleshooting small engines, *100*, 100–101, *101*
two-stage snow throwers, *72*, 72
two-stroke cycle engines, 42–43, *43*

V

valve life, 168
valves, 44, *45*, *46*, *48*, 48
valve service, 162–174
 accessing valve chambers, *164*, 164
 adjusting tappet clearances, 169
 inspecting valves, 166, *167*
 lapping valves, *168*, 168
 machining valves, 164
 replacing valve seat inserts, 168
 servicing Briggs & Stratton Intek® 6 HP single-cylinder overhead valves (OHVs), 170–174, *171*, *173*, *174*
 servicing valves with automotive-type or pin retainers, *166*, 166, 170
 servicing valves with spring (keyhole) retainers, *165*, 165, *169*, 169
valve designs, *162*, 162–163, *163*
valve life, 168
valve spring compressors, 162, *165*, 165
valve spring compressors, *5*, 5, 162, *165*, 165
V belts, 64–66, *65*
venturis, 52, *53*
vibratory plates, 35
Vise-Grip® pliers, 2, *3*

W

walk-behind rotary lawn mowers, 24–27, *25*, *26*, *27*
washing machines, *17*, 17
wiper rings, *50*, 50
wood chippers, *20*, 29, *30*
wrenches, *1*, *2*, 2, *4*, 4, *5*

Z

zero-turn lawn mowers, *27*, 27–28, *28*

Small Engine and Equipment Maintenance

LightUnit Tests 1-5

"God's Light in Electives"

Christian Light Education

Small Engine and Equipment Maintenance

LightUnit 5 Test

Name _____

Date _____

80 / 100   Score _____

Small Engine and Equipment Maintenance Test 5

Circle the letter of the correct answer. (4 points each.) [100]

1. Carbon is the dark-colored soot that can collect on the _____.
    A. cylinder head   B. cylinder wall   C. engine valves   D. all of the above

2. For best cleaning results, soak a cylinder in a bath of _____.
    A. ammonia-based cleaner   C. carburetor cleaner
    B. water and vinegar   D. bleach and soap

3. Battery electrolyte can severely burn the eyes and skin because it is extremely _____.
    A. flammable   B. explosive   C. corrosive   D. liquid

4. _____ shortens the life of an exhaust valve.
    A. combustion   B. overheating   C. death   D. faulty wiring

5. Batteries produce _____ gas, which can cause an explosion if ignited.
    A. helium   B. oxygen   C. methane   D. none of the above

6. A stator has _____ on separate metal stubs, which expose it to a magnetic field.
    A. two electro-magnets   C. a continuous copper wire
    B. carbon copies   D. exhaust valves

7. Automotive-type valve spring retainers are also referred to as _____ retainers.
    A. pin   B. pan   C. pun   D. pen

8. The air gap between the stator and the _____ must be set precisely to function properly.
    A. ignition armature   B. flywheel   C. cylinder head   D. valve face

9. Replacing a failed armature is _____ than repairing it.
    A. better   B. easier   C. harder   D. slower

10. To remove a cylinder head, use a _____ to pry it off.
    A. flathead screwdriver   C. socket wrench
    B. nylon-faced hammer   D. none of the above

11. For best results, the cylinder head should be removed and cleaned after every _____ hours of operation.
    A. 25   B. 50   C. 75   D. 100

Small Engine and Equipment Maintenance

LightUnit 4 Test

Name _____

Date _____

| 80 / 100 | Score _____

Small Engine and Equipment Maintenance Test 4

Circle the letter of the correct answer. (4 points each.) [100]

1. A small engine technician should disassemble a recoil assembly because _____
    A. special skills are needed.
    B. specific care and safety precautions are necessary.
    C. there is a risk of serious injury from flying parts.
    D. all of the above

2. An engine contains a mechanical governor or a _____ governor.
    A. hydrostatic   B. pneumatic   C. secondary   D. polymeric

3. Microns _____
    A. are the smallest part of an engine.
    B. are the unit used to measure filtering material.
    C. are the power in combustion.
    D. none of the above.

4. A tang over the movable post of a _____ prevents it from dislodging.
    A. fence   B. brake band   C. engine   D. muffler

5. How soon should a properly running brake bail stop a running engine after being released?
    A. within 3 seconds   C. immediately
    B. under 1 second   D. none of the above

6. Many fuel tanks use a vented cap to prevent a _____ from forming in the fuel line.
    A. blockage   B. salt   C. surge   D. vacuum

7. To protect against the effects of water, alcohol, and salt what should be done to the fuel tank?
    A. it should be coated with paint.
    B. it should be coated with non-corrosive material.
    C. it should be washed with carburetor cleaner.
    D. it should be brushed with a soft-bristle brush.

8. A rewind starting system is also known as a _____ system.
    A. recoil   B. rope   C. mechanical   D. gear

9. Erratic engine behavior is known as _____.
    A. hunting and surfing   C. hunting and shooting
    B. hunting and seeking   D. hunting and surging

Small Engine and Equipment Maintenance

LightUnit 3 Test

Name _____

Date _____

80 / 100   Score _____

Small Engine and Equipment Maintenance Test 3

Circle the letter of the correct answer. (4 points each.) [100]

1. Use a ____ for final tightening of engine bolts.

    A. torque wrench   B. socket wrench   C. slip-joint pliers   D. leather gloves

2. ____ are devices connected to and powered by a machine for a certain task.

    A. Attachments   C. Implements
    B. Generators   D. None of the above

3. A ____ is used to reduce noise.

    A. foam   B. filter   C. muffler   D. OEM

4. Use a ____ to extend the life of fuel that is not or cannot be stored properly.

    A. fuel filter   B. gas can   C. oil pan   D. fuel stabilizer

5. The first layer of defense an engine has against dirt is ____.

    A. a muffler
    B. an air cleaner
    C. engine degreaser
    D. a soft-bristle brush

6. Drive systems are divided into three basic categories: flexible belt, disc, and ____.

    A. compression   B. alloy   C. friction   D. hydrostatic

7. Use a ____ to test the ignition of a small engine.

    A. breather   B. spark tester   C. lubricant   D. all of the above

8. Which of the following is an attatchment?

    A. tiller   B. backhoe   C. broom   D. snow thrower

9. Debris under the ____ can cause an engine to overheat.

    A. rotary mower deck
    B. stop switch
    C. blower housing
    D. spark plug gauge

10. Flexible belt and disc drive systems can be adequately maintained by ____ with some acquired knowledge.

    A. professionals   C. proper storage
    B. robots   D. nonprofessionals

Small Engine and Equipment Maintenance Test 3

11. If driven by a ____ drive system, a lawn mower can be safely stored without preparation.
    A. hydrostatic   B. manly   C. disc   D. V belt

12. The presence of dirt and debris plays a role in component ____.
    A. power   B. allergies   C. failure   D. none of the above

13. When an engine has leakage around the valves or rings, compression of the air-fuel mixture is ____.
    A. filtered   B. reduced   C. compressed   D. increased

14. A flexible belt must be ____ from time to time.
    A. oiled   B. torqued   C. painted   D. changed

15. The recommendation for changing oil is generally after every ____ hours of operation.
    A. 15   B. 25   C. 35   D. 50

16. The most ____ implement used with outdoor power equipment is a rotary mower deck.
    A. common   B. useful   C. expensive   D. all of the above

17. When an auger strikes a solid object, the ____ fracture, greatly reducing the chance of damage.
    A. snow throwers   C. cooling fins
    B. shear pins   D. engine bolts

18. An operator control switch is used to ensure the ____ of the operator.
    A. comfort   B. power   C. control   D. safety

19. Inconsistent firing (spark miss) can result in ____ engine operation.
    A. improved   B. sluggish   C. extended   D. overheated

20. Which belt is the most commonly used type of flexible belt in the outdoor power equipment industry?
    A. The S belt   B. The T belt   C. The U belt   D. The V belt

21. ____ accumulating between the engine parts can cause a temporary loss of power.
    A. Debris   B. Grass   C. Dirt   D. All of the above

22. A ____ drive system is typically used to transfer power from an engine to an implement or equipment drive train.
    A. mechanical   B. hydrostatic   C. automatic   D. none of the above

3

Small Engine and Equipment Maintenance Test 3

23. If a lawn mower stops unexpectedly while mowing around trees or bushes the ____ may have been accidentally disconnected.

    A. brake B. battery C. stop switch D. tires

24. Fuel containers and tanks should be kept ____ and properly capped to reduce air exposure.

    A. emptied B. filled C. mixed D. none of the above

25. The surface of a ____ will remain hot enough to burn skin shortly after the engine is shut off.

    A. filter B. muffler C. breather D. spark plug

Small Engine and Equipment Maintenance Test 4

10. A properly adjusted _____ can maintain a steady engine speed regardless of tough conditions.
    A. carburetor B. flywheel C. governor D. mechanic

11. A labyrinth fuel tank _____
    A. is easy to get lost in.
    B. has a self-rinse mechanism.
    C. reduces sloshing of fuel.
    D. vaporizes gasoline.

12. Older fuel tanks are made of steel, while modern ones can be made of _____ material.
    A. polyalcohol B. polymeric C. polysulfide D. polyatomic

13. To inspect and test a brake system, _____ and inspect it for nick, cuts, debris, and other damage.
    A. remove the brake pad
    B. remove the brake bail
    C. remove the brake anchor
    D. all of the above

14. On modern small engines, the _____ fins cool the engine by distribution air around the engine block.
    A. fish B. air-conditioning C. carburetor D. flywheel

15. Most carburetor problems are caused by _____
    A. using diesel instead of unleaded gasoline.
    B. driving the mower too fast.
    C. dirt particles, varnish, and other deposits.
    D. none of the above

16. The most common tool used for governor spring adjustment is a _____ tool.
    A. tang bending B. governor wrench C. torque wrench D. spring bolt

17. Only fresh _____ should be used in a small engine.
    A. kerosene B. unleaded gas C. undyed diesel D. water

18. To tighten mounting bolts to 40 in-lb use a _____.
    A. torque wrench B. monkey wrench C. crescent wrench D. socket wrench

19. The brake band should be replaced if the pad's thickness is less than _____".
    A. 0.080 B. 0.075 C. 0.008 D. 0.090

20. The idle screw is always set _____ the governed idle speed.
    A. at half B. less than C. above D. to match

Small Engine and Equipment Maintenance Test 4

21. As engine speed increases the governor gear _____ its rotation.

    A. begins   B. ends   C. speeds   D. slows

22. The fuel pump may be mounted _____.

    A. on the carburetor

    B. near the fuel tank

    C. between the carburetor and fuel tank

    D. all of the above

23. The number of _____ in the fuel filter affects the amount of fuel that can flow.

    A. holes   B. filters   C. screens   D. foams

24. A safe way to inspect or run the engine is _____

    A. tying the brake bail with a bungee cord.

    B. holding the brake bail by hand.

    C. removing the brake bail.

    D. none of the above.

25. Carburetor cleaner is a powerful cleaner that _____ if they are left in a long time.

    A. can clean plastic parts

    B. can strengthen plastic parts

    C. can harm plastic parts

    D. can eliminate plastic parts

Small Engine and Equipment Maintenance Test 5

12. Electrical problems will _____
    A. prevent onboard electrical devices from working.
    B. cause an explosion if ignited with a spark.
    C. deposit carbon buildup in the combustion chamber.
    D. all of the above.

13. Valve _____ repairs minor scoring and pitting of the valve face and seat.
    A. chamber   B. lapping   C. gauging   D. rotators

14. Small engines usually use _____.
    A. lithium-ion batteries   C. lead-acid batteries
    B. alkaline batteries   D. none of the above

15. Some valve assemblies include circular components called _____ that turn the valves slightly to ensure even wear patterns.
    A. spinners   B. friction discs   C. armatures   D. rotators

16. Use a _____ to check the gap between the flywheel and the armature.
    A. feeler gauge   B. stator   C. spark tester   D. socket wrench

17. Use a _____ to check an ignition system before changing a faulty armature.
    A. feeler gauge   B. stator   C. spark tester   D. socket wrench

18. Carbon deposits should be cleaned out by removing the _____.
    A. valve face   B. breaker point   C. drive disc   D. cylinder head

19. Which of the following are a result of carbon deposits in the combustion chamber?
    A. higher oil consumption   C. engine overheating
    B. engine knocking   D. all of the above

20. When slippage occurs in the locomotion of equipment, replace the _____.
    A. friction disc   B. drive disc   C. valve rotator   D. faulty armature

21. Ignition service includes _____.
    A. replacing ignition armatures   C. retrofitting old model ignition armatures
    B. testing stop switches   D. all of the above

22. Use a valve spring compressor to _____.
    A. clean carbon deposits from the valve   C. spring the valve into the piston
    B. remove the valve   D. fill the valve with air

3

Small Engine and Equipment Maintenance Test 5

23. The use of unleaded gasoline _____.
    - A. reduces carbon deposits
    - B. cleans away carbon deposits
    - C. burns up carbon deposits
    - D. all of the above

24. Deposits on a valve are generally formed by _____.
    - A. burning diesel
    - B. burning kerosene
    - C. burning oil
    - D. burning natural gas

25. To verify that the push rods operate the rocker arms the _____ must be manually rotated at least two revolutions.
    - A. mower
    - B. flywheel
    - C. rotator
    - D. cylinder

CLE

"God's truth equipping God's people to do God's work."

CHRISTIAN LIGHT PUBLICATIONS

Small Engine and Equipment Maintenance
LU Tests 1-5 606507